HEART OF THE MACKENZIE

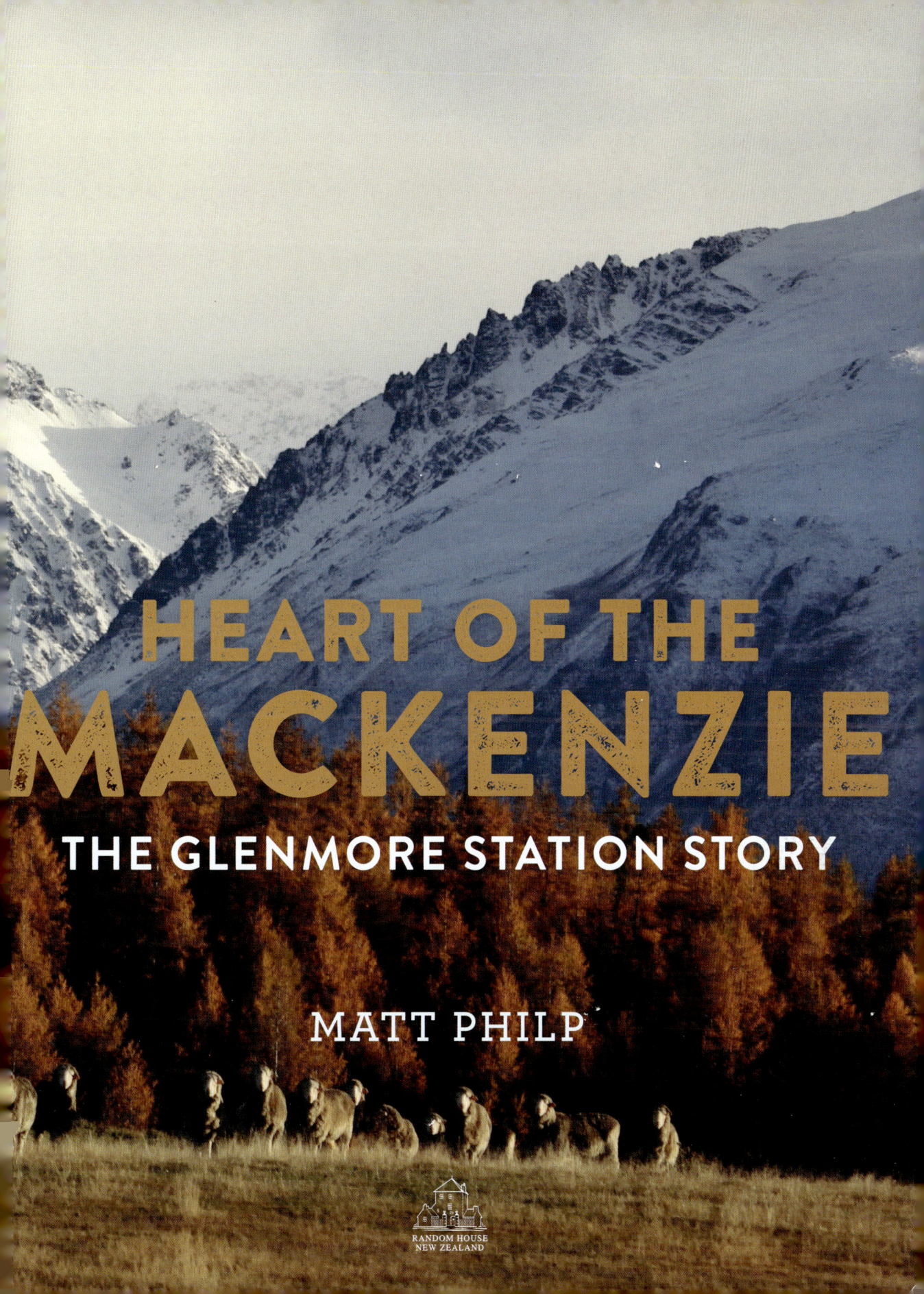

Our original intention for this book was simply to record for our grandchildren — Milly, Harry, Suzie, Angus, Greta and Ben — the stories of their family at Glenmore during the past 100 years. The snapshots of our history must echo those of so many other high country families, nevertheless we hope you all enjoy this book.

Jim and Anne Murray
Wanaka
July 2014

CONTENTS

FOREWORD
12

INTRODUCTION
14

CHAPTER ONE:
GLENMORE'S BEGINNINGS
20

CHAPTER TWO:
THE MACKENZIE COUNTRY STORY
38

CHAPTER THREE:
GERALD'S PASTORAL RUN
52

CHAPTER FOUR:
JIM TAKES OVER
72

CHAPTER FIVE:
GLENMORE — NATURAL AND BUILT
84

The Cass Valley in early winter, with a nor'wester on the way. The jagged peak at centre left is Hell's Gates.

CHAPTER SIX:
JIM LEAVES PASTORAL FARMING BEHIND
110

CHAPTER SEVEN:
THE NEIGHBOURS
140

CHAPTER EIGHT:
WOOL, BREEDING, SHEEP — AND GORDIE
150

CHAPTER NINE:
THE GLENMORE WAY
168

CHAPTER TEN:
PESTS, PLAGUES AND POLITICS
186

CHAPTER ELEVEN:
THE CHILDREN'S EXPERIENCE
196

CHAPTER TWELVE:
HUNTING TAHR AND SAVING STILTS
208

CHAPTER THIRTEEN:
THE FIRE
218

CHAPTER FOURTEEN:
LAKE ALEXANDRINA
234

CHAPTER FIFTEEN:
PREPARING THE WAY
242

CHAPTER SIXTEEN:
THE AUTUMN MUSTER
280

ACKNOWLEDGEMENTS
302

GLENMORE AND THE MACKENZIE COUNTRY

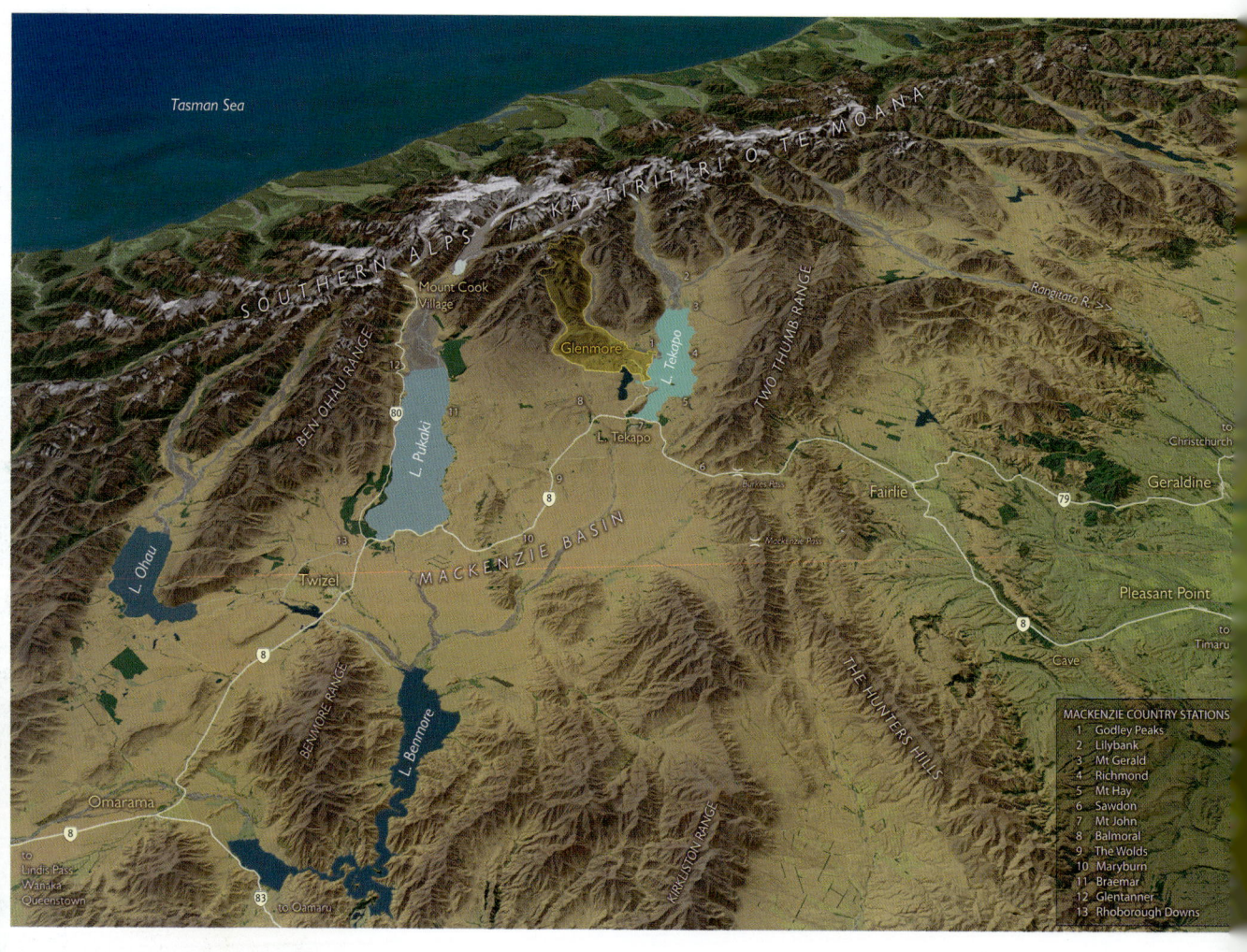

Will and Ems Murray riding over the Glenmore downs. Mistake Hill is behind them, and the Cass River gorge lies beyond the belt of trees.

FOREWORD

As we leave behind the green landscapes of the Canterbury Plains and foothills and enter the Mackenzie Basin (also known as the Mackenzie Country) through Burkes Pass, the changes in natural landforms and colours are dramatic.

Named after the legendary alleged sheep-stealer James McKenzie, the Mackenzie Basin is the largest inter-montane basin in New Zealand, and is renowned for its extensive tawny tussock grasslands enclosed by mountain ranges to the east and the Alpine Divide to the west, culminating in snow-clad Aoraki/Mt Cook, the highest peak in the country and a formidable challenge to mountaineers past and present.

The Mackenzie Basin is a landscape built by the uplift of mountain ranges and their subsequent modelling by glacial processes, resulting in lakes such as Tekapo, Pukaki and Alexandrina, ephemeral tarns and ponds, glacial moraines, and vast areas of outwash surfaces and associated rivers covering its floor. Superimposed on these ancient natural features is the recent network of roads and dams and canals of the huge Waitaki power scheme, the towns of Tekapo and Twizel, airfields and ski fields, tourism facilities, the camp and exercise area of the New Zealand Defence Force, and other notable features such as the Aoraki/Mt Cook National Park and the Mt John Observatory. More recently the spectacular 'skyscape' of the Basin has been recognised globally by the establishment of the Aoraki Mackenzie International Dark Sky Reserve. It is no wonder, then, that the Mackenzie Basin has become a mecca for tourists and other visitors from home and abroad, and the focus of people eager to see its natural features and biodiversity maintained where possible.

The Mackenzie Basin is also about the people who have settled there and drawn a living from the land. The first of these were the itinerant Polynesian hunter-gatherers during the early part of the last millennium, who, through the use of fire, promoted the extension of tussock grasslands at the expense of woody vegetation such as forest and scrub, and used the area to hunt for moa, weka and other products of the land. Evidence of their activities and impact can be seen in the fossil remains of woody vegetation in soils, and in middens and cooking ovens, especially around lake margins. But that is another, albeit significant, story of the Basin's cultural history recounted in our archaeological literature.

The 1850s heralded the beginning of permanent settlement of the Basin by European settlers following the division of the land into Crown leasehold pastoral runs or stations. This book, ably written by Matt Philp, captures the history of one of these runs, Glenmore Station, and its principal guardians, four generations of the Murray family. As such it provides a microcosm of farming and the farming families of the Mackenzie Basin over the past 150 years or more. In it we are reminded of the trials and tribulations faced by farming families of the Basin, including those brought about by isolation, extremes of climate, land tenure, major weeds and animal pests, a procession of regulatory statutes, and confrontation with environmental pressure-groups based within and outside the Basin. At one time the Basin's close-knit and supportive rural community was well served in its farming and related needs by the government departments of agriculture, lands and survey, forestry and the local catchment authority. The subsequent dissolution of these organisations and their practical and resident field staff has created a major void unlikely to be addressed in the future.

On a personal note, it has been a privilege to have been invited by the Murray family to visit Glenmore Station on several occasions to explain some of the outstanding natural values of the property and the practical management of these values within their farming operation; an example being the permanent covenant protecting the iconic Glenmore Tarns, the most outstanding example of their kind in the South Island high country.

Dr Brian Molloy
ONZM CRSNZ AHRIH
Christchurch, 2014

Dog power. Mt Hutton in the background.

INTRODUCTION

The thrown fleece sails through the air, falling on the classing table with a soft *whump* to signal the start of shearing at Glenmore Station in the Mackenzie Country. At the head of the table, Jim Murray scrutinises the wool for a beat, noting with a satisfied nod the absence of stains, raddle and other imperfections.

ABOVE
The shearing team, 2012.

RIGHT
The Glenmore shearing team in full swing.

Alongside him a dozen shed hands, all women and aged from their early twenties to late fifties, wait for instructions on handling this year's wool from Jim, and from Pania Warwick, a slight, self-possessed Maori woman who has been lead shed hand at Glenmore for over a decade.

It's been a year of ups and downs at Glenmore — in other words, a typical year of farming in the high country near Lake Tekapo. A big snow in June jangled nerves, until an unseasonable series of July nor'westers cleared the bulk from roads and paddocks. The year's worst news — wool prices collapsing — was also happily cushioned thanks to Glenmore having a fixed-price contract with New Zealand Merino, suppliers of Icebreaker.

These are perennial concerns of farming life, and they still colour Jim Murray's thinking, although less vividly than they once did. After five decades of farming at Glenmore, Jim and his wife Anne passed the farm to their son Will and his wife Emily ('Ems') before retiring to a new home in Wanaka, two hours' drive south via the Lindis Pass.

That geographical separation has been important. Jim was a schoolboy when his father Gerald had a heart attack and he was thrust into running Glenmore. The first year he lost half his flock to a record-breaking snowstorm. He farmed firstly to save the station, then transformed Glenmore into one of the high country's most successful merino operations, a farm that Icebreaker founder Jeremy Moon would use to showcase his brand to overseas buyers, the kind of property that invites the label 'iconic'.

Glenmore is in Jim Murray's blood, and leaving it has been a struggle. 'You don't live on a place for over sixty years, then walk off one day and flick a switch,' he remarks. But while the farm is now in Will's hands, Jim returns for all the key moments in Glenmore life, including today's shearing, to which he brings years of experience as a classer.

This year's clip needs to be handled carefully. Only the blackest parts of the skirtings need binning — 'just the absolute minimum', Jim tells the shed hands, as Pania demonstrates by tearing off some scrappy edgings. Last year they removed just two bales of first pieces; this year should be similar. At various locations around the shed are bins for a range of impurities — 'blood/pen', 'urine', and so on — all clearly marked and easily accessible.

For the shearing gang, who left Timaru before dawn to make Glenmore by 7.00 a.m., there are some adjustments to make from their last job. 'Each shed is different,' remarks Jim. 'The contract requirements, the type of wool, the country it's coming from all vary, so the gang has to be able to adjust.'

The next couple of hours will be a kind of feeling-out period for everyone. Jim, who runs a tight shed, as organised and efficient as every other part of the Glenmore operation, is preoccupied with making a good start and ensuring that the wool is handled correctly and at an appropriate pace. The half-dozen shearers and Pania's crew will look to find their rhythm, the shed hands working two shifts of four women on the boards, four on the tables, with a break for lunch and a short rest every couple of hours.

Festooning the wall behind them, and hanging from the rafters above their heads, is a mass of pennants, ribbons and certificates. There are wins from the Mackenzie Highland Show. The Temuka/Geraldine A&P Association. NZWTA/Otago Merino Association. The Royal Show. The Canterbury Show. So many honours, all reinforcing Glenmore's history of excellence in producing fine merino wool.

For Will Murray, however, history is less important right now than the state of this year's fleece. It's been a late start to the shearing, with the gang delayed by bad weather, and Glenmore close to the last shed on its itinerary. He's hoping that a new practice of shearing the sheep's bellies in autumn will help to make up time. But his main concern is quality rather than speed. Half of the station's income for the year is bound up in this wool. 'You spend twelve months getting this far,' says Will, 'and it can take five minutes to muck it up.'

Up on the stage, the shearers have finished their preparations. The six orange Sunbeam shearing machines have been oiled, blades checked for sharpness. A couple of the older men have strung up harness belts, like great Jolly Jumpers, over which they will bend to save their backs. Their feet, like the others, are shod in moccasin-like leather slippers. Everyone's ready, just waiting for the sign from Jim.

RIGHT

Glenmore deer in winter.

Lake Alexandrina. Mt John is visible in the centre distance.

CHAPTER ONE
GLENMORE'S BEGINNINGS

Run 79, Glenmore Station, occupies 19,120 hectares at the north-west edge of the Mackenzie Basin, bounded by the Cass and the Fork rivers. A long and narrow property, it begins in cultivated downs beside Lake Tekapo and ends in the alpine barrens and snowy peaks of Aoraki/ Mt Cook National Park, a climb of 1905 metres, with an annual rainfall difference between top and toe of more than 1.2 metres. This is not a farm set in paradise so much as in two different worlds.

It *is* paradisally beautiful, though. Even an official landscape report, normally no place for sentiment, veers at times towards the poetic, cataloguing the 'mosaic of rock and scree' that dominates the station's upper reaches, where glaciation has carved out great U-shaped valleys, and where musterers huts nestled at the base of slopes are totally overwhelmed by the scale of surrounding landforms'. The presence of gem-like tarns among the tussock-clad terraces, the transient nature of the high country light, the play of seasons, the curve of the river valleys, 'which contribute so much of a sense of mystery and intrigue', are all noted in a tone caught somewhere between the inventorial and awestruck.

Who knows what kind of emotional response the landscape struck in Joseph Beswick, who took up a 20,000-acre proto-Glenmore in 1858, but you can bet it was a fleeting one; of far more concern to Beswick was how well the country could grow wool.

Beswick already knew the area well, having spent time working at neighbouring properties Balmoral and The Mistake, then owned by his friends TW and GW Hall. In the official application, the property was referred to simply as 'East, Lake Tekapo'; Beswick named it Gristhorpe, a solid Yorkshire name from a somewhat less than solid Yorkshireman. Tempted by low rents but badly lacking in capital, Beswick failed to meet the stocking conditions of his licence and had to forfeit the property, not once but twice, before hitting more serious financial trouble. On Christmas Eve 1866 he delivered his good friend John Hall an early present: the run and 4000 sheep, gone for a song.

Hall, who came to New Zealand from Hull in the north-east of England, changed the name initially to The Castle, working the run as part of a family property portfolio with his two brothers' neighbouring runs. Between them, the Halls owned 130,000 acres in the Lake Tekapo area, as well as Terrace Station on the Rakaia. More impressively still, John Hall was able to rise through the ranks of provincial and national politics to become premier of New Zealand in 1879, relinquishing the position three years later due to poor health.

Hall took more than a glancing interest in his Tekapo run, stocking it with an additional 3560 sheep, and sacking his manager when most of that flock perished in the winter snows of 1867. According to Robert Pinney's *Early South Canterbury Runs*, the new manager ran the station in close association with Hall's Balmoral run, effectively creating one massive property spanning from the Cass to the Tasman River. The home run equated with the lower part of modern-day Glenmore.

In 1868, Hall flicked his Tekapo station to Alfred Cox, who promptly

RIGHT
Looking up the Cass Valley in winter.

BELOW
Hunting sheep out on Tin Hut Block. Mt Lucia is the high peak at centre left, Memorial Hut is just beyond the belt of pines, and Ailsa Stream is in the middle ground.

LEFT ABOVE

The original old Glenmore homestead by the lake.

LEFT BELOW

Late 1870s: men at the Glenmore woolshed in the days when the station ran 40,000 sheep.

RIGHT ABOVE

Mary Murray (née Gebbie).

RIGHT BELOW

John George Murray, Jim's great-grandfather.

renamed it Castle Hall before selling it on to one John McGregor. In McGregor, Glenmore finally got an owner — and a name — befitting its potential as a high country run of special status.

John McGregor was a teak-hard Highlander, a former boundary keeper at Grampians Station, south-east of Tekapo. He was never devout, but was reliably hardbitten. While at the Grampians he wrote about the daily task of keeping the flock within the unfenced boundaries: 'I had to go over the top of the Grampians Hills every morning and if I were not away by four o'clock I reckoned I had slept in.'

After taking on the property in 1873, McGregor discovered a 5000-acre block in the forks of the Cass River that had not yet been claimed. When neighbouring runs belonging to The Mistake Station were later adjusted, the Scotsman got all parts on the Cass's western bank (the true right), with the river as the boundary. Pinney describes this upland holding as 'the original and true Glenmore', notwithstanding the various leases of the previous two decades. Certainly, it was McGregor who gave the property its name, referencing Glenmore Valley in Banffshire in his native Scotland.

McGregor lacked luck, however. He lost much of his flock in the winter of 1888, and struggled to keep his head above water during a protracted downturn in wool prices. By the early 1890s Glenmore had been taken over by his creditors, the Loan and Mercantile Agency Company, and was being farmed in tandem with nearby Balmoral Station.

In 1910, the land commissioners recorded four deserted homesteads in the Mackenzie

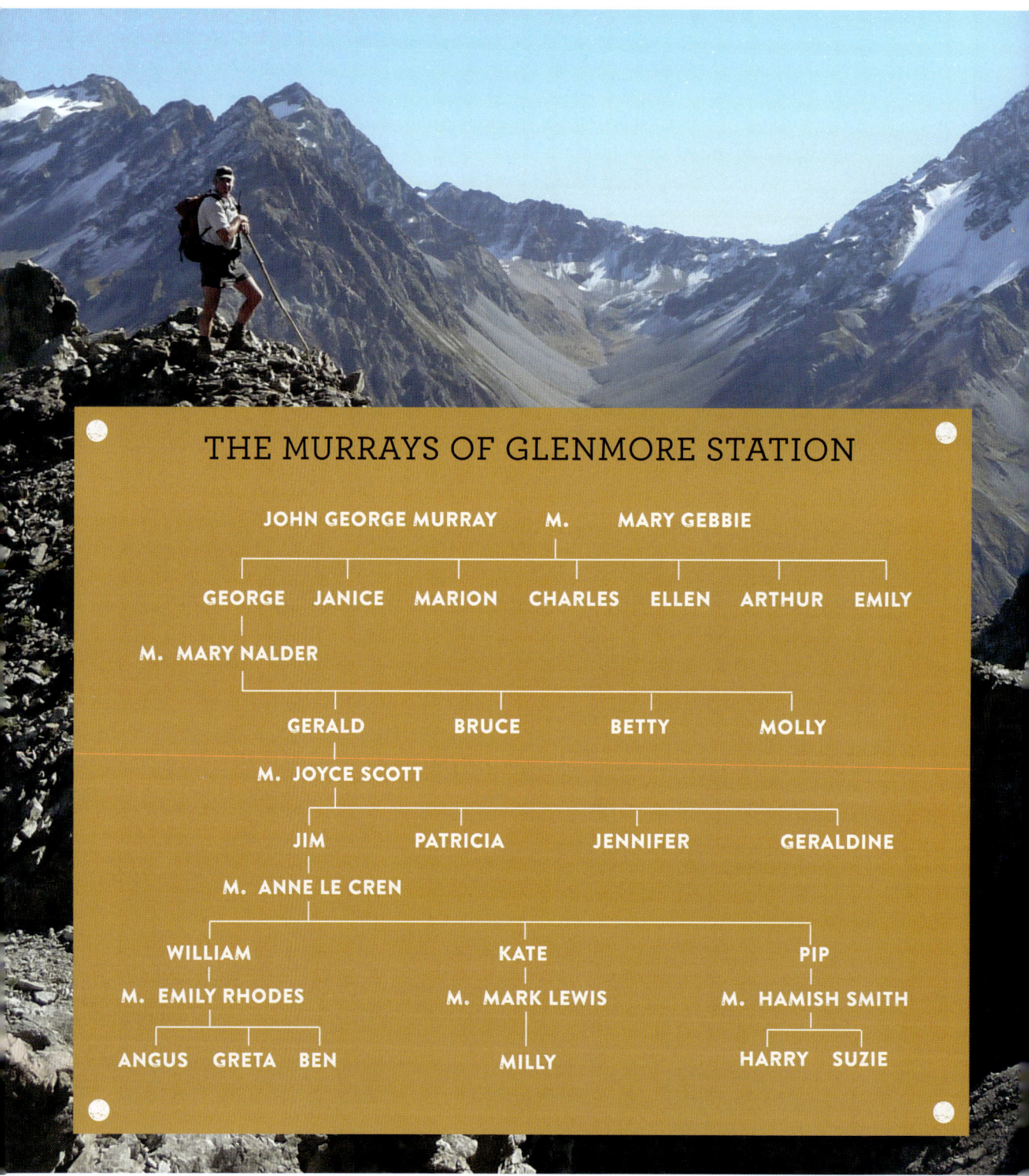

THE MURRAYS OF GLENMORE STATION

JOHN GEORGE MURRAY M. MARY GEBBIE

GEORGE JANICE MARION CHARLES ELLEN ARTHUR EMILY

M. MARY NALDER

GERALD BRUCE BETTY MOLLY

M. JOYCE SCOTT

JIM PATRICIA JENNIFER GERALDINE

M. ANNE LE CREN

WILLIAM KATE PIP
M. EMILY RHODES M. MARK LEWIS M. HAMISH SMITH

ANGUS GRETA BEN MILLY HARRY SUZIE

HEART OF THE MACKENZIE

LEFT (BACKGROUND)

Will Murray looking into Ailsa Stream. Mt Lucia is at the right.

Country, one of them Glenmore. The following year, its boundaries were redrawn, with Mt John removed. Renumbered as Run 79, it was leased to Miss Roma Hope, who two years later sold the lease to Herbert Nalder, Jim's maternal great-grandfather. When Herbert died in 1918, Glenmore passed to his daughter, Mary Murray, wife of George Murray. It would be farmed on Mary's behalf by a manager until one of the Murrays' two sons was old enough to take it on.

Jim's grandparents George and Mary Murray had what you might call the right pedigree for the high country — at least, on George's side. Mary's father Herbert Nalder was a solicitor from Lyttelton. But George's grandparents had been on the first settler ships to reach New Zealand, and brought farming heritage: his grandfather John Gebbie's family had farmed a property in Scotland for three centuries. George's grandmother, Mary Gebbie, was the first European woman to set foot in New Zealand.

George's mother, also Mary Gebbie, was the eldest of John and Mary Gebbie's seven children. His father was John George Murray, a farmer whose parents had arrived on a ship called the *Libuan* in 1851 and settled on a property at Tai Tapu, near Christchurch. George grew up the eldest of seven children at the Murray family farm of Riverlawn.

Both George and Mary Murray were independent, questioning types who broke with family tradition. Resisting his father's plans for his future, in 1894 a twenty-year-old George took off to see the world, working firstly as a foreman in a South African diamond mine. The engagement ring that Jim gave to Anne dates from this trip by his grandfather.

'George was a gutsy bugger,' says Jim. 'After the mines, he went to Argentina, where he managed a sheep and cattle ranch, and then on to England. Later when he was farming in New Zealand he was one of those tough cookies. He had a fob watch, and as soon as you came on the property you were on the clock. But that would have been the way he was brought up working in South Africa and South America.'

Mary, too, marched to the beat of her own drum. A bright and capable woman, she had shown little interest in Christchurch social life, choosing instead to train as a nurse under Sibylla Maude, the legendary founder of New Zealand's district nursing service. At first, Herbert Nalder refused to allow his daughter to work, arguing that she would be taking money from women who needed it. When Mary insisted, he threw her out of the house. Eventually, Nurse Maude persuaded him to relent.

Jim recalls Mary better than he does his grandfather. 'My grandmother was made of good stuff. She was out of bed at six, had a cold bath every morning no matter the weather, worked hard, and was in bed at nine every night.'

Mary and George met in Australia, where Mary had travelled after completing her training, to nurse a friend who had fallen ill. That friend, Frances Rolleston, was at the time hosting her oldest brother, George Murray. In his diary, George noted the encounter: 'Miss Nalder and I bicycled down the hill to the village and spent the afternoon playing tennis.'

LEFT ABOVE
Mary Murray.

LEFT BELOW
George Murray.

RIGHT
George and Mary Murray's headstone in the Tai Tapu cemetery, near Christchurch.

They married at St Michael's Church, Christchurch, on 27 February 1898. The *Press* account described Mary as 'looking exceedingly well in a cream dress of serge. The bodice was of cream silk beautifully tucked with lace. Her large cream straw hat was trimmed with silk bows and plumes.'

Early married life was less glamorous, being characterised by constant movement and the hardships of high country life. Like his grandson, George was highly entrepreneurial, and in 1896 bought the first of several Mackenzie Country properties he would own, Sawdon Station.

Located near Burkes Pass, Sawdon was reputedly among the coldest homesteads in New Zealand. During the blizzard of 1903, George proved both his mettle and his nous by bringing 20,000 sheep to safety using a horse-drawn snowplough he'd invented. Later that year they sold Sawdon and bought Glentanner, a station on the west side of the Tasman River running from Mt Cook to the outlet of Lake Pukaki, paying £9000 to LGD Acland. (Coincidentally, Glentanner features in Anne Murray's own Mackenzie Country story: her great-grandfather Harry Le Cren eventually purchased the station.)

In 1905, George purchased Rhoborough Downs in Mary's name, before offloading it to Donald MacRae three years later. Then, in a high country version of trading up, the Murrays shed Glentanner for Braemar Station, on the opposite side of Pukaki.

The context of the Braemar purchase is revealing of the changing times. A few years earlier, in 1898, all the Mackenzie leases had expired and were put up for auction by the government. But with the tide turning against large-scale landholding, the name of the game now was subdivision. Glentanner, for example, initially a massive holding, was divided into three smaller stations: Top House, Ferintosh and Pukaki Downs.

GLENMORE'S BEGINNINGS   29

Braemar, which under the previous ownership of the Acland family had stretched from the shores of Pukaki to as far as the Fork River, near present-day Glenmore, had likewise since been subdivided. Even so, it was still a large station. Paying £5 an acre, George and Mary bought the Braemar homestead and 6700 acres of surrounding freehold land, to go with close to 60,000 acres of government leasehold land in the Jollie Valley, including all lands around the spring basins to all the country on the true right of the Fork River.

Braemar was no less isolated than any of the other homesteads the Murrays had owned. Fairlie was the nearest town, but still a distant forty-seven miles (seventy-six kilometres) by gravel road. A journey to Timaru, where all important farm business was conducted, involved enough of an ordeal to warrant at least four days at the Grosvenor Hotel, where the Murrays would host friends for afternoon teas and dinners, and generally get a feel for life outside the station.

Mary mitigated the isolation by becoming the first woman in the Mackenzie Country to drive a car. She was resourceful and organised, and like other high country farmers' wives was adept at running a large household with limited contact with the world. When their alcoholic cook went on a bender during shearing, she drove the 100 kilometres to Timaru to find a

ABOVE

Tekapo House under construction.

RIGHT

Gerald Murray and his brother Bruce meeting the Duke of Gloucester at the laying of the foundation stone of the Church of the Good Shepherd in Tekapo. Tekapo House is in the background.

replacement. When none could be found, Mary spent the evening learning to make bread at a Timaru bakery, then drove home to cook meals and bake bread in a brick oven for the thirty-man gang.

George, meanwhile, was cementing his reputation as a hard-nosed and dedicated farmer with a talent for getting the best out of a property — qualities that Jim would inherit. His abilities as a breeder of merino and Romney sheep, Clydesdale horses and Jersey cattle were well known in the district, and he enjoyed great success sheepdog trialling with his imported Border collies, and also running his Scottish Highland cattle on Braemar. In 1934, alongside holding the Braemar and Jollie leases, he purchased a dairy farm at Temuka, where he ran Jersey milking cows.

'By this time Murray had years of high country experience,' writes William Vance in *High Endeavour: The story of the Mackenzie Country* (1965). 'He raised the productive capacity of the station, using up-to-date implements for soil improvement, and raised first class cattle on the Tasman River flats.'

NEIGHBOURING STATIONS

Glenmore has an historical relationship with Braemar by virtue of the Murray family connection. But it is also closely linked geographically to a handful of lakeside merino runs known collectively as the Tekapo stations. Arrayed clockwise around the lake from the Fork River, the sequence is Glenmore, Godley Peaks, Lilybank, Mt Gerald, Richmond and Mt Hay, originally Tekapo Station.

The story goes that Godley Peaks' first owner, Thomas Hall, actually picked out Glenmore on a map as his first choice, but in an apparent confusion over boundaries ended up farming the other side of the Cass River — hence the station's early name Hall's Mistake, later simply The Mistake. Hall worked The Mistake in conjunction with his brothers George and John at Balmoral and Braemar, before selling. Later it was owned by Mackenzie country identity Nicolo 'Big Mick' Radove, a black-bearded Sicilian whose physical exploits included shearing fifty merinos before breakfast. On Radove's watch, The Mistake became infamous for its bibulous week-long parties — perhaps explaining why he went broke. After various changes of ownership, Jim's grandfather George Murray bought the property in 1921, renaming it Godley Peaks, which was run by his son Bruce. The Scott family, whom Jim's family got to know well, farmed Godley Peaks from 1941 until Bruce Scott's death in 1995.

Lilybank, at the head of the lake between the Godley and Macauley rivers, is probably best known to the public from the days in the 1990s when it was owned by Tommy Suharto, son of the corrupt Indonesian dictator, who turned it into a hunting lodge and took an antagonistic approach to public access. The first owner to stock Lilybank with sheep was William Sibbald, a former sea captain who took on the run in 1868 but left most of the farming to a manager. Under Sibbald, Lilybank became famous for its annual horse muster, which according to William Vance drew a big audience and traditionally concluded with exhibitions of horsemanship. In 1911, Mt Gerald was subdivided off Richmond, but they were still seen as one, with Mt Gerald used as wether country. After the 1914–1918 war, Jack Pope took over Mt Gerald and ran it as a separate unit until it was sold to Rex Malthus in 1937. He continued to run it until 1961, when it was sold to John Hogg.

The early history of Richmond Station is dominated by syndicates rather than families. The first partners to take up the run were Messrs Aikman and Wilson, early Canterbury settlers with entrepreneurial blood in their veins. Charles 'Billy' Newton, whom they engaged to manage the station, seems to have also had shares. In the 1880s, brothers Andrew and William Grant bought Richmond in partnership with Arthur Hope, with Hope becoming sole owner in 1890. It was terrible timing: in the great snowstorm that struck five years later, Richmond Station lost all but 300 of its flock of 20,000. Hope rallied, however, and by 1899, when it was sold to James Pringle, Richmond was in the black. Apart from a three-year break in the 1920s, it remained in the Pringle family until 1954, when it was sold to Don Waters, the runholder with whom Jim Murray and two others founded Roundhill ski field on the property.

Tekapo Station no longer exists, but it figures strongly in the pioneering history of the district. John and Barbara Hay, who took up the station in 1857, were the first permanent settlers in the Mackenzie Country. 'Mrs Hay, known as "the Mother of the Mackenzie", was the first white woman to go through Burkes Pass,' writes Vance. 'The Hays are also credited with having the first permanent flock of sheep in the district.' In 1911, the station was subdivided into two blocks. The homestead block was owned by Anne Murray's grandfather Vivian Le Cren from 1914 until 1949, after which it was absorbed into three surrounding properties. The second block, known as Mt Hay, was owned by John Scott for close to twenty years before he bought Godley Peaks.
With thanks to William Vance's High Endeavour.

In 1918, George installed an electrical generating plant at Braemar, the largest of its kind at the time in the Southern Hemisphere. He was also largely responsible for the introduction of hydro power at Fairlie. At a time when electric power was mostly limited to the towns, Braemar had electric radiators in every room and floodlights for the tennis court.

George expected people to give their best — the fob watch was often used — but he could also be generous. Among other things, he paid for the land on which the Church of the Good Shepherd stands at Lake Tekapo. 'He had a community spirit that has gone through the whole Murray family ever since,' says Anne.

One beneficiary was the Barwood family, whom George financed into business as carriers based in Fairlie. This was not charity — someone was needed to cart Braemar's wool and to bring in fencing gear, fertiliser and all the other necessaries of farming life. But when the family became financially strapped, George kept the business running.

Says Anne: 'He went down one day and found old Bill Barwood having to stuff straw into the tyres to keep the truck on the road, so George went off and bought new tyres from Timaru and just gave them to Bill.'

Interviewed at the age of ninety-six, shortly before his death, Bill's son Dougie Barwood recalled George's help: 'George Murray said to my father, "Would you start a transport business? I will supply the trucks for a start and I want you to cart to Braemar." I remember him coming down from Braemar on a Friday. He had a new Model A truck, and as soon as he'd poke his nose in our garage, Dad would say, "George, could you let us have a few bob to buy some petrol?" This was in the slump, you see. And I don't think many months went past where Dad didn't have to approach George Murray. Dad must have paid it all off because later on Dad bought more trucks.

'George was a hard old case sort of a guy. He'd bought a straight-eight Hudson. Crikey! It had a big long bonnet. It was only a three-seater. He'd come down on a Friday for his council meeting and he'd always pull in — and he'd say, "Hullo, young Barwood! I want all these things collected." So I'd get in the Hudson and I'd drive around town to get all his groceries for him. I was only fourteen and I thought, this was great! I never even had a licence!'

In the 1920s George instituted the Braemar Sports, a Scottish-themed *Top Town*–style event, which was held every Boxing Day for the next thirty-odd years. To the sounds of a Highland band, contestants tossed the caber, chased greasy pigs, and took on elaborate obstacle courses, ending the day with the Tasman Valley Handicap, an all-in running race in which people of various ages were ranged the length of a half-kilometre course. More than anything, the day was a chance for neighbouring runholders and their families to get together and relax, away from the daily stresses of their station lives.

———✕———

ABOVE
Gerald Murray.

Jim's father Gerald grew up in this high country world of Braemar Station, with brother Bruce and sisters Molly and Betty. In his early years before boarding school — starting at Waihi Preparatory School, Winchester, followed by Christ's College, Christchurch — it was Braemar that shaped Gerald.

But Gerald ultimately turned his back on the kind of life and style of farming that George favoured. Braemar was an intensive, forward-thinking farming operation for its time. There were two lots of teamsters on the property handling ploughing and cultivation. There were stopwatches and high expectations, and no time for quiet contemplation of mountainscapes.

Anne has a phrase she uses to describe Jim's early zeal, recalling that from his first days in charge of Glenmore 'he was going to break eggs with a big stick'. The expression is apposite for Jim's grandfather, but not for Gerald. 'I think my father vowed and declared that he would never cultivate a paddock, or have a teamster on his property,' says Jim. 'He wanted a traditional pastoral run, something very low-cost, with none of the hassles of producing supplementary feed. It reflected his approach to life. He enjoyed other things in life than farming. But that's human nature, isn't it? You get a generation that is very driven, and the next generation says, "No, I'm going to button back."'

At the age of twenty-six, Gerald was given a choice that would define the lives of his descendants. His brother Bruce Murray had already been installed at Godley Peaks Station in 1921, while George was still running Braemar. (In 1941 Godley Peaks was sold to John Scott, and Bruce returned to run Braemar until his death in the late 1950s, after which the station was sold.

GLENMORE'S BEGINNINGS

Bruce's sons Michael and Tim went on to purchase Mackenzie properties The Wolds and Maryburn, which their respective sons John and Martin still farm.) Now Gerald's father was offering him a choice of two runs. One was Ben Avon, located in the Ahuriri Valley on the far western edge of the Mackenzie. It was beautiful country, idyllic in the summertime, but the winters up the valley were known to be brutal.

Gerald considered his second option with great care. Glenmore, the neighbouring gorge property that his mother had inherited all those years ago, was being run by a manager, but it was Gerald's if he wanted it. He decided that he very much did, perhaps swayed by the comfort of being nestled between family on both sides.

In 1926, the tall and laconic Gerald installed himself at his new property. He may have been less driven than his father, but he had his own plans. Glenmore was a beautiful pastoral run still being managed by traditional methods. And, by and large, Gerald would stick to that age-old script — a man, a couple of dogs and a mighty, unfenced expanse of Mackenzie high country.

RIGHT

Bringing sheep down from Top Block and heading them towards Tin Hut, where they will be held for the night before being moved towards the homestead paddock.

Looking back down the Cass.

CHAPTER TWO
THE MACKENZIE COUNTRY STORY

Let's not dwell on the sheep-stealer too long, other than to say that James McKenzie was at once the antithesis and the embodiment of the country that took his name (under an adjusted spelling).

The antithesis because he stole rather than slogged, skulked rather than stood tall, shot through rather than stayed — the settlers who followed McKenzie into the basin that he discovered in 1855 were by contrast industrious and obdurate.

The embodiment? Because he was a raw-boned Scotsman at home in foul weather and in his own company — rugged and lonesome — and because he could handle sheep like a necromancer.

Sheep made the Mackenzie Country, just as they made the rest of New Zealand. The difference was that in the Mackenzie they were scattered across vast high country runs; merino sheep, for the most part, with a sense for footing in the tussock and scree of the tops.

The harshness of the country selected the Mackenzie's human inhabitants every bit as effectively as its sheep. With the best farming land in Canterbury already taken, the Mackenzie Country appealed as a lost world — a promised land, even. But the Englishmen who took up the first runs didn't last long, having little experience of mountain sheep farming and no temperament for the solitary life. Properties in the Mackenzie changed hands frequently, a winnowing that over time filled the place with Highland shepherds, a breed far better suited to its fierce demands.

'Heredity and environment had conditioned these shepherds to little cob huts, to loneliness, and to fight it out,' writes Vance in *High Endeavour*. 'They were prepared to "endure all things" to make a home for their children, and for their children's children.'

The Mackenzie would have evoked memories of home. To the east, the rampart of the Dalgety, Kirkliston and Two Thumb ranges, covered in snow and beset by fogs. West and north-west, the Ben Ohau and Gamack ranges, prelude to the greater bulk of the Southern Alps, presided over by Aoraki/Mt Cook. Within the basin, arranged north to south like a bird's claw, the two glacier-fed lakes Tekapo and Pukaki and Lakes Ohau and Benmore, all part of the Waitaki catchment.

It is a landscape of glaciation, of U-shaped valleys and cirques sculpted by the advance and retreat of ice. In a

RIGHT

The view from Tin Hut across the Cass River and on to the Haszard Range.

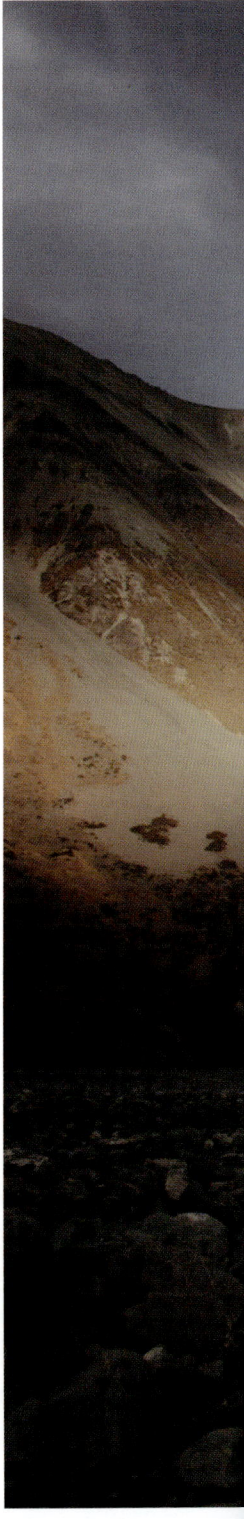

LEFT

A stream below Waterfall Hut. The hut is the small white dot at the base of the hill in the distance at left.

guest chapter for *High Endeavour*, geologist LE Oborn writes that the most common features of the Mackenzie geology are the lateral, median and terminal moraines and outwash gravel plains. 'The moraines, those hummocky dumps of rocks, were built of detritus tipped off the advancing ice, or were left by it when it retreated. The vast plains of outwash gravel, in places up to 1000 feet thick, were deposited by the streams issuing from the glaciers.'

Whatever bush existed was long gone by the time the first European settlers arrived in the Mackenzie. In its place were vast billowing grasslands, studded with matagouri and a few other stubborn, low-growing shrubs, blue-grey in colour and with well-developed root systems capable of searching deep for moisture.

What is it about this country that gets to people? Anne Murray, whose father's early childhood and that of his siblings was spent at Tekapo Station, recalls how her uncle returned to the Mackenzie a few years before he died. 'He came over Burkes Pass and tears began flowing down his face, and he turned to me and said, "I feel I'm home." Yet he'd lived almost his entire life in England.'

The magic is in the emptiness, that strong sense of being in a basin under a big sky, a kind of desertscape broken infrequently by a few shelter pines or a vivid azure-turquoise lake. For some, it's a landscape that doesn't immediately inspire fond feelings. Evelyn Hosken and her husband William travelled from Ashburton to take up a run at Simons Hill in 1911, planning to occupy it on behalf of her brother for five years. In a memoir, *Life on a Five Pound Note*, she describes feeling dismayed on arrival at Tekapo. 'I wondered how I could ever live on that barren tussock country for five years.'

For earlier runholders, dismay wasn't a sentiment they could afford. The Mackenzie was their big chance to make a life of sheep farming in New Zealand, if not a fortune. After 1851, when restrictions on run sizes were lifted, there'd been a mad rush to take up property in Canterbury; by the time of James McKenzie's capture in 1855, very little in the province was left unoccupied. As news trickled out of potentially vast sheep country beyond the 'Snowy Mountains', the most intrepid of land seekers started taking up Mackenzie runs.

Vance describes the process. 'When a settler discovered suitable land not already taken, he made a rough map of the area, then hurried back to the Lands Office, Christchurch, to apply for a licence to occupy it. Provided the land had not been previously applied for, the Land Board granted this licence on condition that the applicant stocked the run with one sheep to every twenty acres, or one head of cattle to every hundred and twenty acres, within six months.'

North-facing country was prized, but there wasn't a great deal of it — the Mackenzie mostly turned its back to the winter sun. The plains and the downs went first, followed by more mountainous country only after it was demonstrated that sheep could be run there. Wherever people settled, the country's light soils ruled out any serious attempt at cultivation. The Mackenzie would be built on large-scale pastoral farming — on wool — because it was the only viable option.

In *Early South Canterbury Runs*, Robert Pinney remarks that leasehold was adopted from the start as the best vehicle for opening up the country and generating quick revenue. But the cash-strapped pioneers who took up these huge runs on very low rents immediately encountered problems stocking the land to the requirements of their licence. Sheep at the time were in short supply and prices were high. Runholders were forced to look far afield, often all the way to Nelson, and getting their new stock to the Mackenzie was a trial. Tussock country was very often burned to make the way easier for driving sheep.

Vance quotes John Hall on his efforts to overland a mob to his Tekapo run: 'There were no convenient fences or friendly paddocks where sheep could be secured, and no homestead or even sheds where you could obtain a night's shelter; the weather had to be faced, whatever it was; wild dogs were occasional visitors, so that somebody had to watch all night, and if a sou'wester came tearing across the plains, all hands would have to stand out in it to prevent the mob from breaking away.'

RIGHT

Nor'west arch.

Weather was ever the bane of a Mackenzie Country farmer. The summers are dry, bakingly hot and cursed with wild and oppressive nor'westerly winds. Merinos can handle that. Less forgiving are the bitterly long winters and legendary snowfalls. By common consent, 1895 was the worst.

The great snow of 1895 was in fact not one storm but many, a succession of falls that arrived as early as April and continued through the next few months, producing the severest winter ever experienced in the South Island. In the Mackenzie the snow was compounded by a series of frosts, writes Vance, 'that not only prevented the snow from melting, but also froze the surface into a fine dust, which, blown by the wind, formed a frozen mist that clung to clothing and hair in miniature icicles. Heavy lakeside fogs clouded out the sun in the Mackenzie Basin . . . 1895 is remembered mostly because it wiped out about half the flock of almost every station.'

In the South Island's high country, three-quarters of a million sheep were killed. At Glenmore, where Lake Alexandrina froze so thick that cattle were able to walk across, most of the flock was lost — some 40,000 out of a total of 45,000 sheep shared between Glenmore and neighbouring Balmoral.

Not only sheep perished. At Richmond Station, where twelve feet of snow fell and some linen hung out in April remained snow-buried for five months, twenty-eight dogs were frozen to death in their kennels, and horses and cattle were also lost.

John Rutherford, who owned The Mistake (now Godley Peaks) in 1895, wrote up his midwinter visit to the station for the *Timaru Herald*. His impression of the Mackenzie was of an unbroken sheet of white except for a few clear strips on a handful of stations, where surviving sheep huddled. 'I was advised to go up by boat, as there had been no one down from either Glenmore or The Mistake except by boat . . . the splash of water froze on the oars instantly. All along the beaches of Lake Tekapo and Alexandrina there were sheep in numbers, alive and dead on the ice, some frozen fast by the wool. Glenmore hands were out skinning every day and had skinned 700 already. On The Mistake, all the sheep that have been seen out of 15,500 were a few on a hillside near the homestead, and very few of them were alive. . . . Those that are alive are so wasted that one can lift them by the back like a kitten.'

Jim Murray's first year running Glenmore under his own ownership was 1967, when a destructive spring snowfall dumped 132 centimetres over the Mackenzie, and much deeper than that in drifts, disrupting transport and communications and causing heavy sheep losses.

Will and Emily Murray took over Glenmore in 2006. The next winter, Canterbury was hit by a record snowfall that closed most South Island rail

MEMORIALS OF THE MACKENZIE COUNTRY

James McKenzie's memorial requires an effort to find, being miles away from the main highway on the western side of the pass he discovered. On a slab of local bluestone is inscribed: 'In this spot James MacKenzie [sic], the freebooter, was captured by John Sidebottom and the Maoris "Taiko" and "Seventeen" and escaped from them the same night, March 4, 1855.' Sidebottom was the overseer at The Levels (coincidentally, owned by Ems Murray's great-great-great-grandfather George Rhodes); Taiko and Seventeen were two Maori shepherds. After capturing McKenzie as he was about to settle down for the night, they attempted to herd both prisoner and sheep back up 'this awful hill', only to lose the Scotsman in thick fog.

At the top of Burkes Pass is another bluestone monument, this one honouring Michael John Burke, who discovered the pass in 1855. Burkes Pass the settlement was the early social and business hub for the Mackenzie, but withered when the promised railway line stopped short at Fairlie. The memorial to Burke was erected by TD Burnett, MP and owner of Mt Cook Station, who was a lifelong advocate of tree planting in the high country. 'Plant forest trees for your life,' the inscription advises those entering the Mackenzie Country.

Near Cave, the Herbert Hall–designed Church of St David is named for the patron saint of shepherds, and includes stained-glass windows dedicated to the pioneer women of the Mackenzie, 'who through Arctic winters . . . maintained their homes and kept the faith'. On a brass tablet nearby is inscribed a dedication to Andrew and Catherine Burnett, TD's parents, who took up the Mt Cook run in 1864 'and in the wilderness, founded a home'.

lines and Canterbury schools, toppled trees and cut power.

'We had a metre of snow around the homestead for six days and there was no heating except for a log burner, so I was sleeping in the kitchen with my six-week-old baby Angus beside me,' says Ems. 'There was no power, no road access for a week, and the air temperature was so cold that although the freezers weren't working everything remained frozen. The fire sprinkler system in the garage was shattered by the cold. It had dripped, and an icicle had formed from the ceiling to the top of the vehicle. The dishwasher froze in the kitchen and blew out the pipes. I remember taking my new baby out for a walk one day and it was still minus nine, but you just had to get out in the fresh air.'

If there is such a thing as a distinctive Mackenzie Country culture, it draws deeply on the experience of such a harsh environment. Thomas Burnett, owner of Mt Cook Station and an MP, wrote in 1909 of the killing effect of the worst snows, remarking that 'with all these losses it takes a man who is fairly oozing optimism to believe that in time the Mackenzie will be a much more liveable place for sheep'. Vance emphasises the individualism and stoicism of the pioneers — they just got on with things, as do their descendants.

A different insight comes from Michèle Dominy, an American professor of anthropology whose research for *Calling the Station Home: Place and Identity in New Zealand's High Country* took her on a

RIGHT
Climbing out during the autumn muster, with the Faraday Glacier in the background.

THE MACKENZIE COUNTRY STORY

Glenmore autumn muster. She suggests that high country inhabitants resist the romanticisation of their landscape and way of life.

'The lessees who shaped my understanding taught me to see landscape not solely as scenery . . . but equally as inhabited by and embodied in stock and humans. I learned to look at a hillside with an eye for the pinpricks of sheep and cattle, for faint traces of fence lines and tracks, for a musterer's potential beat.'

But where a runholder sees working country, others see 'iconic' landscape and tourism potential, or a fragile ecosystem demanding protection. The Mackenzie is changing fast, and consensus on whether it's for good or bad is as hard to spot as merino on a tussocked slope.

Tourism is becoming more important to the district's economy — worth $96 million in 2012, much of it thanks to being on the increasingly busy route between Christchurch and Queenstown. Recent recognition of the Mackenzie Country as an international dark sky reserve is expected to provide another boost. Importantly, though tourism's star is rising, sheep farming in the Mackenzie high country has become an increasingly marginal operation, stung by fast-rising costs while returns have stagnated. Most controversially of all, the move to intensive irrigation and dairying that has transformed the Canterbury Plains has now crossed Burkes Pass.

The appearance of paddocks of green alongside the highway between Omarama and Twizel is as confronting as it can be when it comes to landscape modification in the South Island. Many who know and love the Mackenzie will struggle with it, even if they accept the rationale. Others can find no argument to justify what the Royal Forest and Bird Protection Society has referred to as 'this creeping green stain'.

Meanwhile, in the long-running tenure review process, some lessees such as Glenmore are considering surrendering some of their high country land to the Crown in return for freeholding other areas. Will Murray is probably voicing the opinion of many runholders when he says that the Mackenzie is beautiful country to farm, but also terrible country to farm, because so many people feel they have a vested interest in what happens there. 'If you want to be left alone, well, you can't be,' he says.

It sounds like a well-rehearsed complaint — but it's not. Will is of a new generation of runholders getting to grips with a very different farming environment from the one their fathers knew. At Glenmore, he and Ems have instituted a robust farm management plan that includes regular monitoring of water quality, tussock cover and other significant natural values — grabbing the initiative rather than having change foisted on them.

Elsewhere, too, there is a push underway to shape a consensus. In 2013, something called the Mackenzie Agreement was unveiled, a document aimed at bringing together environmentalists, farmers, tourism operators and others traditionally at loggerheads to forge a way forward for the district. It's a work in progress, of still unknown value. Labour says it sells out the environment to development. Forest and Bird has cautiously welcomed the agreement, but says the devil will be in the detail.

The agreement allows for ten per cent of the basin to be irrigated, and accepts limited intensive cubicle-style dairying on five properties. Against that, it plans for enhanced control of wilding pines and the restoration of native vegetation in some areas, and it envisages a trust being established to protect water quality from the adverse effects of land-use intensification.

The future of the Mackenzie Country is very much up in the air; in other words, up in that big cerulean-blue sky. Who knows what Will's children will face if they become the fifth generation of Murrays to run Glenmore. Perhaps the farming of merino sheep will be a museum exhibit by then, or a tourism sideshow, but you have to doubt it. The Murrays, as will become obvious, have high country farming in their bloodstream. And sustaining this family, at the very heart of both their economic and emotional life, is Glenmore Station.

Sheep being driven down the Cass riverbed on the autumn muster.

CHAPTER THREE
GERALD'S PASTORAL RUN

It's a simple business, pastoral farming — and a precarious one. An inventory of Gerald Murray's operation at Glenmore in the first years would barely cover a cigarette packet: a couple of horses and dogs, a pair of helpers, and 6000 sheep.

No fences — at least, not in the beginning. No tractor. No supplementary feed; whatever grew in the weeks between the snow melt and the summer had to sustain Glenmore's sheep for the year. This was low-input farming, with minimal costs but an equally low threshold for bad luck and dire events.

'He was looked upon as someone who was incredibly lucky to live in that environment,' says Jim of his father. 'The station just quietly ticked away, because there was no rush to do anything, no great pressure. You can imagine the income off 6000 sheep in those days, when your running costs were close to zilch. But he was no different to a lot of his type in that era — that was pastoral farming.'

A slightly aloof character, not particularly gregarious but in possession of a dry wit and mad keen on his polo, Gerald might once have been described as a 'gentleman farmer'. That phrase misses something of his devotion to the property, however.

'He had a tremendous love for the high country, for the land, and for the wool,' remarks Jim. 'He'd be happy to sit in a chair and just look at the hills all day, watch the changing light. And he loved nothing more than to yarn about the high country and its history.'

The art of pastoral management was straightforward enough: just keep the sheep out of the snow, and make sure the country had enough time to regenerate. There was no irrigation used, no growing of green pasture or secondary feed; Glenmore's 6000 sheep and sixty breeding cows had to survive on the station's God-given covering of tussock and other native grasses, supplemented by 400 bales of hay brought in for winter. Hoggets were run on a 5000-acre block, and Glenmore wintered an additional 3000 ewes between Mt Joseph and Waterfall Hut up the Cass Valley.

For his first dozen years of farming, Gerald didn't realise that the 10,000 acres of the Fork Valley was part of the Glenmore lease, so it was left ungrazed. Once he realised his mistake, he immediately put it to use to graze dry stock during the summer months.

It was a big, unfenced property, one of the largest runs around, and for the most part farmed on horseback. Jim, born in 1944, remembers his father's team of four draught horses and three station horses, and the daily routine of feeding each of them a full bucket of chaff. For years the station maintained a fully equipped blacksmith's shop complete with forge and bellows.

Except for the occasional team of men brought in to help with the muster and for shearing and crutching, Gerald ran Glenmore with only a shepherd and a boundary keeper, the latter a vital part of the operation because of the absence of permanent fencing.

Since the earliest years, boundary keepers had been a staple of Mackenzie Country farming. They were usually Highlanders, rugged, self-reliant types, adept at reading the weather and working with stock, good with dogs and happy in their own company. As Vance describes in *High Endeavour*, they usually lived alone in a cob hut by a stream, or in a dark gorge near the station's borders, spending their days walking the boundary to check for wandering sheep. During winter months, when the snow acted as a natural boundary, the boundary keeper worked with a horse and dogs to make sure the sheep stayed on the sunny faces and ridges.

Weeks could pass without them seeing another soul. Some boundary keepers became so shy of the outside world they would hide in the hills to avoid the packman bringing in supplies. Mostly they lived on mutton, which they slaughtered and butchered, liberally washed down with whisky.

There was a kind of lonely romance to the life — at least, as viewed from the outside. Mackenzie Country balladeer Ernie Slow set one of his best-known yarns, 'The Godley Ghost', in the boundary keeper's world — at Glenmore's Sardine Hut, as it happens, near the Fork River.

Written in 1936, the ballad begins with a strong nor'wester kicking up during the night, causing the door handle to rub against the iron walls with a terrible screeching. Unnerved, boundary keeper Jack Skinner looks out into the night and sees the Devil's daughter 'among the rocks and water'. He shoots, then rides for his life through the storm, pursued by this terrifying shade.

Sardine Hut — so named because it measured just six feet by six feet, with a five-foot lean-to roof — was one of two boundary keeper's huts on Glenmore, the other being at the Joseph yards. In later years the hut was relocated close to the homestead and used as a skin-drying shed.

The boundary keepers' time at Glenmore ended with Gerald's decision to put in permanent fence lines. Made of West Coast silver pine, these fences had to be robust enough to survive conditions at altitudes as high as 1760 metres. They were carted up by packhorse and erected by contract fencers working in sometimes atrocious weather.

There was a knack to building snow fences. Metal standards were of no use, because once the metal had buckled under snow and been fixed a couple of times it tended to break. Instead, wire was lightly stapled to the posts, which were spaced widely enough to mitigate snow building up and doing even more damage. Fences were designed to fall over rather than break under the weight of snow.

When Gerald took over Glenmore the sheep were mostly Romney half-breeds. He made the decisive call to switch breeds, calculating that merinos

THE GODLEY GHOST
by Ernie Slow

Jack Skinner alone can surely boast
Of having seen the Godley Ghost;
'Twas way up in the Sardine Hut,
Where spooks and phantoms nightly strut,
For there, among the rocks and water,
He saw the famous Devil's daughter.

Now spooks, like fleas, they fear the light,
So to the hut they came at night.
Jack Skinner had just arrived, you see,
Back from Fairlie, on the spree.
And like all men, his happiness grew,
With some of Scotland's famous dew.

He danced with glee upon the floor,
When came a tap upon the door:
A visitor at this time of day?
A shepherd must have lost his way,
And open wide Jack pulled the door,
Then staggered back with an awful roar.

He seized his gun for open slaughter,
There before him stood the Devil's daughter;
A female form before him stood,
Jack aimed and fired as quick as he could.
With smoking gun and failing light,
I'm sure he looked an awful sight.

'I've done for her,' he madly yelled.
His chest with pride and spirit swelled;
'All spooks I'll fight; all forms and sizes.
With whisky good, my courage rises.'
But the night wore on with wind and rain,
When a tap came on the window pane.

Standing there, mid'st falling water,
Jack saw once more the Devil's daughter;
Two loud reports, a mighty crash,
That sent the window pane and sash.
And Skinner sank with eyeballs red,
Upon his old and trusty bed.

He prayed the Lord would send the light,
To end this most distressful night;
He stirred the fire, more light to keep,
And went to bed but not to sleep.
While resting on his cosy bed,
The wall was rapped above his head.

'I've been a John Hop in the Force,
I've steered erratic in life's course!
I've taken mad men to the cells,
I've flirted with pretty belles;
But never such a night I've spent,
With nerves and spirits badly rent.'

Upon the hillside, cold and bare,
He saddled up the old grey mare;
'It's for my life I'll ride this race,'
He called for Phar Lap's mighty pace.
Pity poor Skinner in his plight,
As he rode out into the starless night.

He dashed o'er rocks, through scrub and water,
But following fast came the Devil's daughter;
His spurs sank in like spearing fish,
His whip came down with an awful swish;
And from the mane right to the tail,
He rode for life — he couldn't fail.

With speed to burn, I'd not reward her,
But old Jack Skinner was working her harder;
He spied the lake of bluish water,
When upon his back sprang the Devil's daughter.
He called for help — he called afar,
He called for Hamilton's racing car.

The sheep, they scrambled up the rocks,
And wild birds flew away in flocks;
And birds that never flew before,
Flapped their wings, as off they tore.
He jumped the well-known station gate,
'Twas six foot high — he couldn't wait.

Dog kennels upset, and sheep dogs, too,
Flew at the sound of the hullabaloo;
And crashing through the door, half shut,
He galloped into the shepherd's hut.
Dave Sutherland shouted 'Earthquake! Fire!'
And out he dashed in his night attire.

The old mare's head through the window came.
For Skinner kept riding, might and main;
A crash of timber, an awful shout,
The old mare is through; the wall is out.
Dave Sutherland yelled out, 'Damn his eye,'
As the hut, it reeled, and then capsized.

Into the swamp and out again,
Jack wheeled his mare for the Glenmore plain;
Bruce Murray jumped out of his cosy bed,
'Sounds like an earthquake here,' he said.
The mules and horses madly fled,
The bull stood fair upon his head.

The Skinner made for the river water,
Racing for life from Devil's Daughter.
Once more he galloped for the station light,
And now he looked an awful sight.
His eyes they glared like balls of fire,
His hair stood up like fencing wire.

His moustache would clean a twelve-inch gun,
For Skinner then commenced to run;
He flattened the henhouse midst jolts and jars,
The roosters fled right to the stars.
With a mighty effort and plain sweat,
He upset the squatter's dining set.

Around the house, on a beaten track,
He went so fast he saw his back;
The shepherds rushed, but held aloof,
As Skinner climbed up on homestead roof,
As game as Kelly, and riding yet;
They hauled him down with a fishing net.

Then falling at the squatter's side,
The Devil's daughter he defied;
'Oh thank the Lord,' he madly raved.
'Oh, thank the Lord, for I am saved.'
For Skinner didn't care a jot,
For he gulped down whisky, piping hot.

Shepherds still swear, up in the snow,
You can hear those phantom roosters crow;
And travellers, as they pass that way,
Hear them crowing night and day.
And o'er the mountains, rocks and pools,
Three weeks were spent to find the mules.

The bull was found, all stiff and sore,
Just nineteen miles this side of Gore.
Some horses alas were never found,
Some say they're in the phantom pound.
So the boss gave out the following rules,
To shepherds, rabbiters, dogs and fools.

Employees make note and fear,
For whisky is forbidden here;
For months of snow creates less slaughter,
Than a visit from the Devil's daughter.
But they say that whisky often leaks,
Upon the well-known Godley Peaks.

But they mix it well with sparkling water,
To keep away the Devil's daughter.

were better suited to the rigours of the high country, with a fleece less prone to picking up sand and shingle. The decision would be the making of Glenmore in Jim's day, but Gerald never got to experience the full potential of his merinos. Given the complete absence of any supplementary feed, the wool-cut-per-head under his operation never topped two-and-a-half kilos.

While pastoral farming ran to a more sedate pace, the seasonal sheep work during Gerald's day was much the same as now. As today, the autumn muster took place in mid-April and the sheep were crutched in early May, before being put to the ram. Post-muster, the entire flock was put through a swim dip to control ticks, lice and scrapie. Come winter, with few fenced blocks on Glenmore, Gerald's main work involved hunting stray sheep down off the high tops to avoid snows, and ensuring they were well spread through the tussock country.

In October, the ewes were brought down to the lower country to deliver their lambs. For the next several weeks he and his shepherd would quietly ride around the paddocks to check the sheep. While merinos rarely need help with lambing, with ten months of wool on them there was always the potential for them to become cast. Tailing took place in December at the homestead or the Joseph yards.

Shearing was done in two hits, November and January. After branding, the dry stock were turned back up the Cass Valley for summer grazing. Lambs were weaned following January's shearing and ewes culled, with those that had reached the end of their useful purpose driven to the sheep sales.

The Tekapo sheep sales were held religiously on the third Saturday of February. The only merino sale in the South Island, they dominated the late summer. Gerald had a special role here, being one of the instigators behind the building of the saleyards on the eastern side of the lake township. (Prior to 1927, the yards at Mt John Station had been used.) On the Friday before the sales, Gerald, his shepherd and half a dozen dogs drove the lambs and other culled sheep twenty kilometres along the dusty gravel road from Glenmore to the yards, where some 25,000 sheep were penned.

Although Glenmore was run on the proceeds of the annual wool cheque, the sale of surplus stock was an important supplementary earner. In a flock of 6000 sheep, Gerald would retain 1400-odd lambs to go into the flock, leaving on average 400 additional lambs surplus to requirements. It wasn't a windfall by any means — the going price for a lamb was one shilling and sixpence, with up to five shillings paid for a ewe — but it was still income worth having.

The atmosphere of the sales was worth the price of admission, in any case. Set beside the shores of the lake, surrounded by still snow-streaked

RIGHT

Gerald Murray's entries in the Glenmore Station book.

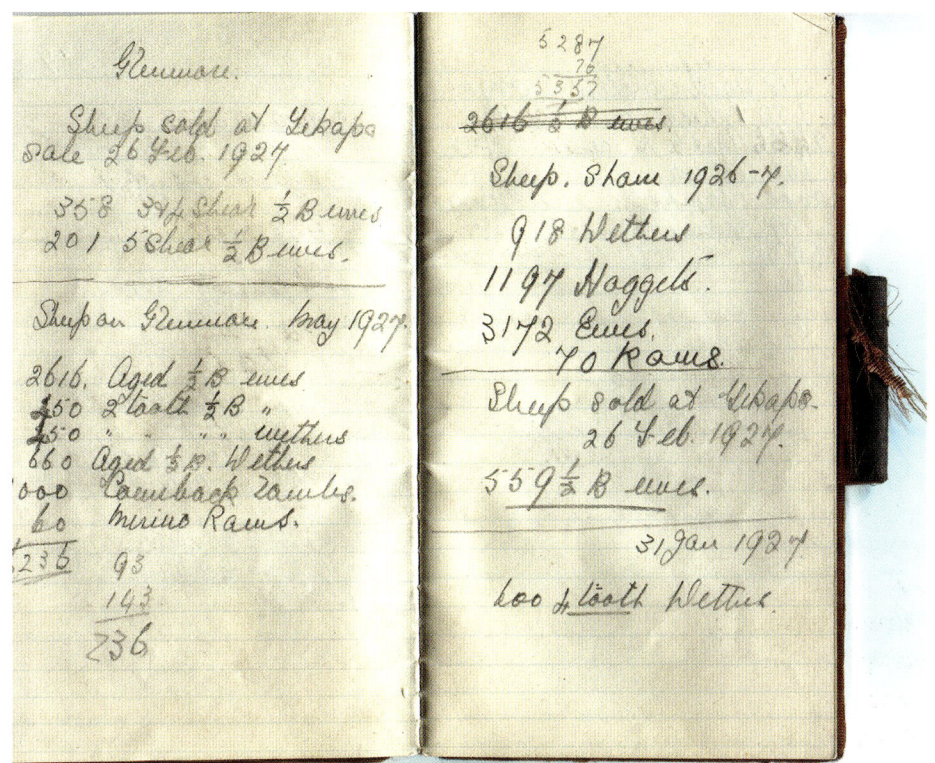

mountains, the saleyards were the centre of Mackenzie life for a weekend. Every farming family in the district would travel to Tekapo to watch the sales and enjoy a picnic, and buyers came from far and wide. Children chased each other around the dusty sheep pens, then raced into the lake for a quick cool-down, while their parents took advantage of the stock firms' generosity with the beer or snoozed under the lakeside pines.

One of Jim's favourite anecdotes about his father relates to the Tekapo sales — albeit tangentially. Gerald's brother-in-law Dick Beauchamp, who had the Mt Edward lease on the other side of Tekapo, was a former British Naval intelligence officer; a charming man, but vague. 'He rang my father one day and said, "I've lost some sheep, can you lend a hand?" They spent three days searching high and low on Mt Edward for these damned lost sheep. On the third day, they were having lunch looking out over Lake Tekapo when Dick disappeared behind a bush. When he came back he said in his refined English voice, "You know, Gerald, I've been thinking about these sheep, and it's just dawned on me while I was having a quick tutai back there — I think I sold the blasted things at the Tekapo sales." At which my father gave him a sharp look and replied, "Bloody shame you didn't have that shit three days ago, Dick!"'

SELLING WOOL

The London Wool Exchange, where Gerald's wool was sold at auction, was very much a world of its own. The following description is from the catalogue entry of a BBC recording of a wool auction in 1935. 'Each morning while the sale season is in progress bales of wool are on show at London warehouses. In the afternoon the auction sales take place in the Exchange, which holds about 400 people. It resembles a large lecture hall built around a sunken platform, which forms the rostrum. Tiers of seats slope upward from the rostrum. From 3 p.m. when bidding begins, the "floor" presents an amazing scene. Buyers of many nationalities join in a babel of tongues that would utterly confuse anyone unacquainted with the wool industry. Bids are made and accepted with a rapidity unequalled anywhere else.'

A 1931 *Daily Mail* newspaper feature took a more colourful licence. 'The scene is a large room which suggests a science lecture theatre. The steep declivity of wooden seats is filled with men. But they are not quiet students. From them burst forth noises like the barking of sea lions; the staccato yelps of guardsmen "numbering off"; the cries of dismay of those suddenly realising intense personal tragedy. The lots go with astonishing swiftness. The real explosions come with the much-desired lots, when the buyers are straining on the leash.'

LEFT

The Glenmore homestead in 1935.

ABOVE

Gerald's REO truck loaded with chaff for the horses, 1936.

Before the introduction of road transport, any stock from the sales bound for Central Otago or Marlborough would be driven down to Fairlie. There they were loaded on a train put on especially for the Tekapo sales and transported to Timaru, where the carriages were separated for Blenheim and Ranfurly.

Glenmore's wool, too, went to Fairlie, taken firstly by dray to the nearby wool landing that Gerald shared with his brother Bruce of Godley Peaks, and from there by traction engine and wagon to the Fairlie wool scour. Once at the scour, the wool was hand-washed in great steaming bowls of hot water and detergent, then pitchforked into rinsing bowls, dried on racks in the sun and pressed.

After scouring, Glenmore's wool was exported through Port Timaru, bound for the famous Wool Exchange, which at the time was located in Coleman Street, near Moorgate in London (see box at left). The shipping, the classing and the sale itself were all handled for farmers by an organisation called the London Wool Brokers.

In 1936, Gerald joined the motorised age, paying £400 for a brand-new REO Speedwagon. (In later years he purchased a Land Rover, a Farmall 'M' tractor with a massive hydraulic loader, and a flat-deck 1944 Hupmobile to cart the cream cans to Tekapo.) A predecessor of the modern pickup truck, the Speedwagon got heavy use transporting wool from all three

GERALD'S PASTORAL RUN   61

ABOVE
The Glenmore homestead in the 1940s.

of the Murray men's properties: Glenmore, Godley Peaks and Braemar. A load of twenty bales encased in rough jute packs went down to Fairlie, while on the return leg the driver was under orders to stop at the Burkes Pass pub to fill a five-gallon jar of beer. A firm tradition had been established that there must always be two beer jars on Glenmore at shearing time: one for the truck, the other at the disposal of the shearers.

Running the station in a more hands-off style meant that Gerald was able to find time for off-farm pursuits. He built a polo field beside the Godley Peaks road by clearing rocks off what is now known as Jimmy's Lagoon, using cairns to mark the sidelines. During the season there'd be a weekly match on this home field, and Gerald also travelled down-country for his polo. (Fearing that wartime petrol rationing might put a crimp in his polo field exploits, Gerald stockpiled a couple of dozen 44-gallon steel petrol drums and stashed them in a tussock gully near the Joseph.) With friends WD Orbell and HH Elworthy he set up the Pareora Polo Club at Timaru, which staged matches for twenty years until disbanding after the Second World War.

They sound like halcyon days, but as Pinney notes in *Early South Canterbury Runs*, 'No high country man is ever complacent.' Under a pastoral farming regime, the difference between a good year and a bad one was decided largely by the spring rains. If October and early November proved dry, to be followed inevitably by an arid Mackenzie summer, Gerald could count on no more than four weeks of growth. There were no drenches at the time to control internal parasites, so along with avalanches, snowfall and starvation, the sheep were at risk of succumbing to basic health problems. During an average winter, Gerald lost ten per cent of his flock.

And there were scourges other than weather to give him sleepless nights. The Mackenzie Country was periodically visited by plagues of rabbits and, as is detailed in a later chapter, Glenmore sometimes struggled to cope. Erosion was another worry.

Although no fan of intensive development, Gerald wasn't without entrepreneurial fire in the belly. When in the 1950s the government instigated a massive project of dam building in the Mackenzie, he spied a chance to create some supplementary income. Reasoning that the mobs of men arriving at Tekapo to build the first dams would need feeding, Gerald launched a butchering operation in partnership with his neighbour Gould Hunter-Weston of Mt John Station, centred on a shop on the road near Mt John homestead. While the other station slaughtered and butchered the sheep, Glenmore took care of the cattle, transporting half-carcasses to the shop on the back of the Speedwagon for processing by a resident butcher.

The foray into butchery was not without its horrors. Worst was an incident involving two young Dutchmen whom Gerald had sponsored to work on Glenmore. Wartime rationing had made its mark on this pair, as became clear one time when they helped Gerald to slaughter a cattle beast. In the moment following the animal's evisceration, the Dutchmen swooped on the heart and liver and began eating. It was too much for Gerald, who made haste to the far side of the yard to throw up.

The new meat venture was never more than a diversion for Gerald. Wool paid the bills, and it was the riddle of how a property could produce better wool that drove him. He worked at it so assiduously, in fact, that he became known as the 'Wool Baron of the Mackenzie' for the quality of Glenmore's wool and the prices it fetched. His son wouldn't inherit many of his personality traits, nor his take on farming, but Gerald's love of stock and his fascination with wool were transferred into Jim like high country DNA.

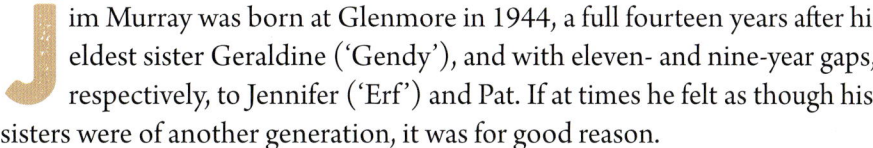

Jim Murray was born at Glenmore in 1944, a full fourteen years after his eldest sister Geraldine ('Gendy'), and with eleven- and nine-year gaps, respectively, to Jennifer ('Erf') and Pat. If at times he felt as though his sisters were of another generation, it was for good reason.

The fact that his sisters were all sent away shortly before his birth reveals something else of the family dynamic. In the world of Gerald Murray's Glenmore, it was considered inappropriate for girls to be too aware of the facts of life.

Gerald and his wife Joyce were in some respects oddly matched. He was taciturn, uncomfortable in crowds, happiest with a mountain view or a history book. Gendy could recall him sitting for hours staring up the Cass Valley. Yet that image of a remote and undemonstrative father doesn't quite catch the man. When his children left Glenmore to return to boarding school, Gerald would shut himself away in the petrol shed so no one could see how upset he was. Another quirk: Gerald loved ships and routinely visited the port when in Timaru on Glenmore business. 'He was invited on board once, a ship loading Glenmore wool to be sold through the London Wool Brokers, and there and then they offered him a passage to London,' says Jim. 'He said he had to turn it down because it was too short notice.'

ABOVE

Joyce and Gerald Murray.

RIGHT

Gerald with a cattle beast, waiting for the butcher's shop.

Finally, there was a disconnect between the tall, powerfully framed man of the land, and his weak heart. By the time Jim hit sixteen, his father would suffer a second cardiac arrest.

Joyce was the social animal. Raised in a well-off South Canterbury farming family and educated at the Dunedin private Presbyterian girls' school Columba College, she never let Glenmore's isolation get in the way of an active social life, inviting a houseful of friends one weekend, putting on a lavish Sunday dinner for extended family the next. Gerald accepted the hordes with good-natured resignation, but escaped to the sidelines as soon as an opportunity arose.

During the pre-war years, Glenmore was a haven of gracious living. Despite the rougher roads of the day, the Murrays went to church every Sunday and routinely had post-service visitors back for a formal lunch. There were tennis parties and a swimming pool, while Glenmore afternoon teas were sumptuous affairs — as Anne discovered after marrying Jim.

'Glenmore was always a place to visit, a family hub. I inherited the remnants of this lifestyle with all these old aunts and uncles coming to visit on Sundays. The tradition quickly faded on discovering that Jim's new bride actually wasn't up to scratch with her Sunday entertaining,' she says.

This life of relative ease was maintained by liberal use of hired help. Joyce had governesses to help raise and school the three girls, who ate with their parents just once during the week at Sunday lunch, receiving the rest of their meals on a tray in the nursery. A married couple, Mr and Mrs Gould, were employed as cowman/gardener and housekeeper, and accommodated in a back room of the homestead. After the Goulds left, Alice Proudlock became cook and there followed a succession of girls to handle the housework.

Although they dined and lived separately, these staff featured strongly in the children's world, and the more colourful are remembered still. There was Bill Bird, a thin, tall shepherd who left Glenmore only once each calendar year, and bathed less frequently. Alice used to send loitering children scattering from the kitchen with a flick of a tea towel and a hissed 'shoo flies!' Among the governesses, Catherine Craddock was particularly prized as a caregiver, a storyteller teacher who on occasion took the girls with her on holidays to Le Bons Bay.

Less fondly remembered was Mrs Robertson. The seamstress visited

ABOVE

The Murray siblings in 1946. From left: Jim, Pat, Erf and Gendy.

Glenmore every year to make clothes for the children. Though handy enough on a treadle sewing machine, she was a chain-smoker and her effusions permeated everything she made.

A feature of the rural life of the day was the work-shy swagger, who wandered roads with a rolled swag at his shoulder, cadging meals from farmers and sleeping rough. Glenmore often got a warning call from the Scott family at neighbouring Godley Peaks if a swagger was in the area, and the children would immediately be packed off. The more convincing of these tramps would occasionally be given a meal before being sent on his way.

For all the privileges the Murrays enjoyed in the pre-war years, there were hardships to living in such a remote and raw part of the country. Because of their isolation, they and their neighbours had to be largely self-sufficient, with any stores brought in just once a month by dray. Leaving the station, too, involved an effort. From Glenmore all the way to Timaru the road was unpaved shingle, a rough day's driving for the vehicles of the day.

In winter, the rutted route around the Glenmore side of Tekapo could be hazardous. Gendy recalled an incident when she and Joyce became stuck in snow. Huddling in the car, they finally heard a horse's hooves approaching — it was Gerald come to find them. Spring, when the frozen ground began to

ABOVE
Gerald at Memorial Hut.

thaw, was almost as tricky, with cars often becoming bogged in mud near the Mt John turn-off. In either season, anyone intending to visit Glenmore was instructed to ring the station from Tekapo before setting out; if they hadn't arrived within forty minutes, Gerald went looking.

Winter was a sterner test than now. The house wasn't insulated to anything like today's standards, and this was a climate cold enough for a cow carcass to be hung in the yards through winter without fear of it going bad. The homestead was kept warm by open fires, and the nursery was heated by a Broadway coke burner, on which the children toasted supper crumpets. Taps were kept running to prevent pipes from bursting. Visitors, by contrast, dried up as the days shortened, and the family's isolation became acute.

In such times the phone became more of a lifeline than ever. Like all rural families of the day, the Murrays were on a party line, with a dedicated ring of long-short-long. Faults were not uncommon, particularly during rough weather, and, as the sheep station designated responsible for maintenance, Glenmore would have to send out a rider to check the line.

By contrast, power was rarely an issue. Inspired by his own father's micro-hydro scheme at Braemar, in the 1930s Gerald installed a direct current (DC) four-kilowatt hydro plant, drawing water from the Cass River and discharging it down a race into Scott's Creek. Glenmore remained off the grid until 1963, according to Jim. 'The only time we got any trouble was when the river flooded, or during an especially cold winter when the whole race used to freeze up. But when that happened we had a standby generator.'

The girls had several outside chores to perform each week. It was their job, for example, to separate the milk. The dairy, a converted horse float, became so cold in winter that they had to warm the milk before they could begin to work with it. Among the sisters, Pat was the most enamoured of farm work, and volunteered whenever she saw an opportunity to help in the yards.

Another chore was fruit-picking. Although it has the ring of a favoured high country ritual, the annual school holidays task of picking Kentish cherries, complete with a community picnic, was not well loved by the children — or at least, never for long. As the day heated up, they invariably snuck away to the lake for a swim and to wash the juice from their arms.

By the time Jim was born, life at Glenmore had become more demanding for everyone. With the outbreak of war, housekeepers and cooks had become unavailable, and Joyce was now doing everything herself, helped by the girls. They washed clothes in an agitator machine filled by a hose from the tap, then manually fed them through a wringer before rinsing and wringing them again. During winter, everything was dried in the kitchen on pulley rails.

Gerald had attempted to enlist, but had been refused on health grounds — bad feet. 'He was issued with a .303 for the Home Guard,' says Jim, 'and honed his shooting skills with it hunting bull tahr.'

Beyond the farm gate, the district was also experiencing changes; some subtle, others more significant. In 1938 the Public Works Department started construction of a tunnel and powerhouse at Lake Tekapo, although it wasn't until 1951 that the station was finally commissioned, marking

LEFT

Jim in 1952.

ABOVE

Gerald, Jim and Joyce.

the advent of large-scale hydroelectricity in the Mackenzie Country. The landscape would soon be bisected by ice-blue canals.

In the early 1940s, the formation of the High Country Committee and the emerging soil conservation movement marked the first stirrings of people other than runholders and the government demanding a say in the management of the high country. Aerial top-dressing and rabbit poisoning arrived at the end of the decade, along with a new Land Act that gave runholders greater certainty on leases.

Back at Glenmore, Jim's upbringing was proving markedly different from that of his sisters, with no governess or nanny on the scene. As a baby he was looked after by a Karitane nurse, but largely it was Gendy who watched over him.

She had her hands full. Among various exploits that earned Jim the label of a troublesome child, he cracked a four-gallon tin of preserved eggs into flour bins and threw the car keys into the creek. He swung the cat by its neck with a rubber band, then let it loose among the chooks. Discipline was meted out by way of a wooden jam spoon, often applied by his mother or by Gendy with sufficient force that it broke across his backside.

Glenmore's distance from the bright lights meant the children had to make their own fun. After a snow, they spent hours sliding down Pudding Hill at the back of the house on a waxed-bottomed tin tray. And in warmer weather when the ranger released fish fry to stock Lake Alexandrina, the Murray children would patrol the creek attempting to spear as many of the larger trout as they could.

Jim remembers swapping farm eggs for trout with the same ranger, Freddie Greyburn. 'I'd go down to the creek with a sugar sack over my shoulder and meet Freddie, who was the ranger for the Acclimatisation Society. They used to trap fish in the creek and strip them of eggs. I can still see him saying, "And what have you come for, young Murray?" and I'd just stand looking at him, and the next minute, "Well, here you go!" and he'd flick a fish to me and I'd put it in the bag and go back to the house pleased as punch.'

Organised entertainment included regular skating parties at Tasman Downs or at the Tekapo ice rink, where the highlight was to ride on 'Bluebird', the ice groomer driven by Mr Rodman. When cinema arrived in Tekapo in the 1940s, the Murrays began attending Saturday night screenings in the tin shed cinema. Most other evenings were spent listening to shows such as the Australian drama series *Dad and Dave from Snake Gully* on the radio, or winding up the gramophone.

The high country year was punctuated by a handful of community and farming events. The Tekapo Sales was the big one for local farming families, but the Fairlie Show ran it a close second.

LEFT TOP

Autumn muster, early 1950s, using the Farmall tractor to tow a cart, supplies and men.

LEFT BELOW

Tekapo House.

BELOW

Joyce and Gerald Murray, 1966.

For Jim, nothing could beat roaming the farm. 'I had a little white pony called Toby I used to ride everywhere. Eventually I went mustering with my father, who had a horse at that time called Diamond. He would put a rope around Toby's neck and I'd sit in the saddle, and away we'd go.'

That close connection to Glenmore's farm life ended abruptly, however, when Jim turned six.

His sisters had all done their first few years of schooling at home, then later at Godley Peaks, riding their ponies across the Cass River every day to school, where they took lessons with that station's four Scott children. Gendy was sent to Craighead boarding school in Timaru aged ten, Erf and Pat at a similar age. (They travelled the three or four hours to Timaru on the back of the REO Speedwagon, always arriving at school for the new term looking very dishevelled.)

Jim, however, only got to do a year of correspondence before he was abruptly packed off to board at Tekapo, followed two years later by Waihi prep near Temuka, and finally Christ's College. 'My mother got sick of trying to deal with me, so I was sent to board,' he offers, only half-jokingly.

Whatever the rationale, from then on Jim's experience of his beloved Glenmore would be limited to his school holidays, right up until the day he returned permanently to help his ailing father to farm it.

GERALD'S PASTORAL RUN

Jim watches as the 3000-head mob heads down the valley on the last day of the autumn muster.

CHAPTER FOUR
JIM TAKES OVER

From a vantage point high above Glenmore, Anne points out a patch of green where Will is trialling lupins as a grazing option. Innovation has been a theme at Glenmore since Jim's earliest days, and Will has inherited that drive. The son wants to stamp his mark on the property. Question is, how much has Jim left to be done?

'Jim is so hard-working and quite entrepreneurial,' remarks Anne. 'Will would often say, "Jeepers, Dad, what's going to be left for me to do?"'

It was a very different scenario when Jim took over Glenmore. For a start, there had been almost no discussion with Gerald about the station's future — even whether there'd be a place there for Jim.

Gerald had suffered his second serious heart attack when Jim was still at boarding school. He was largely an invalid when Jim got the call to come home to help out. He was sixteen, a schoolboy thrust into a farming career while his peers were enjoying their post-college freedom.

'He had no youth,' says Anne, contrasting Jim's early life with Will's. 'He never went to Lincoln, he never got to go away as many of his mates did, bumming around and travelling. His nose was immediately to the grindstone.'

For Jim, however, the path to take was obvious. 'I looked at both of my parents — my father was sixty and my mother in her late fifties — and I realised that if Glenmore was going to stay in the family then I'd better hunker down and make a go of the place. Apart from some mustering on neighbouring places, I was on the property from the day I left school.'

Gerald was directing farm activities from his easy chair with help from his right-hand man Jack Holdom. Jim resolved to learn as much as he could from Jack, as well as from Jim Calder, who worked part-time.

In the summer of 1960, Glenmore was run much the same as it had been in Gerald's early years. There were now two vehicles — the Land Rover and the Farmall tractor — but most work was still done on horseback. At five every morning, Jim would saddle up and ride with his dogs to check the sheep up the Cass Valley, carrying his sandwiches and a flask in leather saddlebags. At the head of the valley he'd spend a few hours pushing sheep further on to the tops, then head home, riding thirty miles to arrive saddle-sore and weary around eight in the evening.

Today, on graded roads and using a four-wheel-drive, the same task takes less than three hours. But Jim has an abiding fondness for the old ways. 'I feel lucky to have experienced the farming techniques of that era. It was a brilliant lifestyle, the pure pastoral run. A horse is still the best way to check sheep, because you can glide through them quietly without disturbing them, and because you're up higher you can look at your country, see what the stock are doing and get a damn good feel for what's happing in the environment.'

He was particularly attached to his horse Blackie, a former trotter with an unusual gait. 'He used to amble — somewhere between a walk and a trot — and it was the most beautiful and comfortable way of riding, and a fantastic way to see the country and inspect stock. Riding home, Blackie

ABOVE
The boundary keeper's hut at the Joseph in the early 1950s.

BELOW
Ewes heading to summer country in the 1950s.

picked up the pace; he knew exactly where we were heading, and my God we covered the country!'

Jim was learning his trade, finding the rhythms of pastoral farming, when Gerald died, aged sixty-five, of his third heart attack.

It was the day of the Fairlie Show, and Gerald had insisted that the rest of the family leave him behind at Glenmore. They got the news at the show — a voice over the PA system told them to hurry home.

Jim regrets now that he didn't ask Gerald more. 'The editor of the *Timaru Herald* said to me not long after my father died, "I don't know why we never got a recorder in front of him to get his stories of the high country and life in Tekapo and at Glenmore." He had such a tremendous recall of history and life as it went on here. I think even in his generation there probably weren't many who'd spent all their life pretty much on one place in the Mackenzie.'

There was another, far more pertinent question that had not been asked. In the six years since Jim had returned to help on the farm, there'd been no serious discussion between the two about what would happen to Glenmore in the event of Gerald's death.

'They were a reserved family,' notes Anne. 'Huge senses of humour, but you didn't talk about any sensitive subjects, you just got on with things. Jim has reacted to that childhood by being incredibly forthright with our own children about where we are all going and what we might be able to do, but as a young man he never addressed the nitty-gritty of things with Gerald.'

But now big decisions had to be made. Glenmore had passed into Gerald's estate — thankfully with no debt, because of Gerald's conservative stewardship. None of Jim's sisters was interested in taking on the farm. Gendy was now married to a North Canterbury farmer, Erf had become a schoolteacher, Pat was a Karitane nurse. The only prospect of Glenmore remaining under a Murray was if Jim could buy out his siblings — and he was barely twenty-one.

It was at that juncture that Gendy's husband, Peter Northcote, stepped

JIM TAKES OVER

into the discussions. Jim says he will always be grateful. 'Peter was at those meetings to give me encouragement and to persuade the rest of the family that I should be given the opportunity to buy it, even though I was so young.'

This was no cold-eyed business decision for any of the parties. And in the end, it was the family's emotional attachment to Glenmore that clinched it.

'There's a sense of belonging that every family in the high country has for the station where they were brought up, and that feeling stays with you until the day you leave this world,' says Jim. 'No matter how old you get, you still have a great respect and love for that land. My sisters quietly thought, great, it's staying in the family. It's still our home.'

Glenmore was Jim's if he could raise the capital. He had no money of his own, having been paid only a nominal wage for his station work. With the support of his lawyer, he borrowed £50,000 from the bank, and in March 1967, aged just twenty-two, Jim Murray became the owner of Glenmore Station.

ABOVE
Gerald.

RIGHT
Jim.

It was a huge responsibility for one so young — 'Financially, I was taking on the world', he remarks. But Jim also saw a tremendous opportunity. Even at that stage, he wasn't interested in just holding the line.

'Whatever industry you're in, a young person coming in wants to make his mark,' he says. 'You're not just going to follow what your father did for the next forty years. You want to leave your footprint on the property, to leave the land in a better state than when you took it on, and the farm in a healthier financial position. That's what farming is all about, after all — the love of the land and to provide for your family.'

On 1 April 1967, he began farming in his own right. He had 5700 sheep, 120 head of cattle, and an income for the property of £22,000, £18,000 of it from the wool clip.

That first winter he scratched through, taking advice from more experienced farmers when he could. 'I remember I couldn't get the ewes up the Cass to settle down on the bottom block. There were two-and-a-half thousand of them, and they kept walking up and down, up and down. My neighbour John Scott told me, "Don't worry, Jim. We've got a full moon at the moment. Wait for a few days until that moon disintegrates and it gets really dark, and they will soon settle and get into some good grazing patterns." And he was so right.'

Joyce had moved to Timaru, so Jim was alone in the homestead, cooking and fending for himself. In the early hours of a November morning he woke to a silence somehow heavier than other mornings — eerie, even. Outside, the land all the way to the horizon had vanished under an expanse of deep snow.

It was 16 November 1967, a landmark date in the Mackenzie, marking one of the very worst snow events of the years since July 1945. This time, however, the snow had arrived in spring, when the lambs were on the ground and the sheep back in the higher country.

Around the Mackenzie Country crops were flattened and flocks decimated. At Glenmore, like many other parts of the district, the snow was well over a metre deep. And when the snow stopped, the rain began, falling almost unrelentingly for a week — seven inches in total. By the time it broke, Jim would be coming to terms with the scale of Glenmore's losses.

In those first few hours after waking, however, he struggled to formulate a plan of action. Then, a knock at the door: shepherd Jack Holdom was outside, having floundered through the snow for two hours from his cottage just 400 metres away. Between them the two men managed to undo the frozen bolts on the shed door and retrieve the crawler tractor. Working with numb hands, they attached a snowplough, then set out to survey the damage.

In the homestead paddocks, all that could be seen of the sheep were their black noses poking out of the snow. The men cleared a path, conscious of the need to move quickly to reach the hundreds more ewes and newborn lambs in the tussock blocks beyond. For all his youth, Jim knew what had to be done.

'In any snow like that, even in the

wintertime, the number one thing is to find where your stock are. You have to get them onto a clear track you've made with the snowplough, then make sure they have access to sunny country. It has to be sloping into the north, because that country is going to clear of snow first.'

The theory was sound, but two men couldn't rescue the flock without help. Unfortunately, most of the central South Island was struggling under the same snow, and in many places power and phone lines were down. Using a radio telephone, Jim managed to contact property owners in the Rakaia Gorge to pass word to the outside world that Glenmore was in trouble.

Within two days the army had sent a helicopter and Jim and Jack were able to survey the entire property. The following day, volunteers coordinated by the Ministry of Agriculture descended on Glenmore in crawler tractors.

'They stayed in the homestead with us and most brought their own food. My sister Erf happened to be staying with me, and she organised the domestic side of things. In the kitchen at night, there would be a gang of men and a couple of dogs warming themselves beside the oil burner. We were without power for ten days. No stores came in. We ate porridge twice a day because we'd run out of bread. It was porridge and spuds that kept us going.'

The volunteers were a godsend. Many were farmers themselves, some from down-country near Fairlie, where conditions were equally bad. 'They were in the middle of lambing, but they kept turning up, saying, "What can we do?" The hardest thing was

RIGHT

Memorial Hut, Top Block. Mt Lucia at left, Mt Hutton at right.

HEART OF THE MACKENZIE

to tell them what I wanted done, because none of them knew the place and we had no clear country to put sheep on. The hay barn had collapsed, so we couldn't even get at what little hay I had.'

Glenmore's neighbours were also struggling to cope. The morning before the snow, Jim had brought in a Caterpillar D6 to help with river protection work up the Cass Valley, and had left it parked a couple of kilometres from the house. On the first afternoon after the snow, Jim heard it splutter into life — Bruce Scott had crossed the river on horseback to requisition the machine for Godley Peaks. Jim didn't mind. 'We all had problems and that's the way things operated back then. If you needed something, you didn't worry about asking — you got on and used it.'

As the snow melted, the extent of the disaster was revealed. Jim found 3000 lambs and sheep dead — half his flock. The survivors were desperately thin. Those few who still had lambs were barely able to feed them and, looking at the rest, Jim figured he'd be lucky to get ten per cent lambing.

Yet metabolically they weren't too bad, a testament to the famed hardiness of merino sheep. Jim was also grateful that they'd been shorn before the weather hit; on some properties, the weight of wet wool in the snow had caused even greater devastation. Among the cattle, losses were less severe. The Angus, being black, were less affected by snow burn than Herefords would have been, although their udders and teats still suffered badly from burning and blistering caused by snow glare. As a result, many of the cows wouldn't let their calves drink, leading to shocking malnourishment.

Confronted by such an appalling scene, it was hard to know where to start. Jim and the volunteers made a dump for the carcasses they could reach; those animals that had died in the less accessible reaches of Glenmore were left to decompose.

It was ugly work, heartbreaking for Jim, and it all took a toll physically. For a week following the snow, he couldn't eat properly and had to drink through a straw because his lips and face were so swollen from the snow glare.

LEFT

They could only be merinos.

RIGHT

Climbing out of snow on Tin Hut Block.

The body would heal, eventually. But could Glenmore survive? 'It took me seven years to farm my way out of that,' says Jim, who rates the '67 snow as having the single biggest impact on his farming life. 'I was just getting started and it clobbered me.'

He went to the local county council for rates remission, then to the Lands and Survey Department for a pause on rent for the pastoral lease.

'I hunkered down. I didn't waste money going anywhere. I stayed on Glenmore for three, four months at a time. In terms of farming, I just stuck to the fundamentals. You feed your stock as best you can — which wasn't easy given there was no hay or anything else on the place. And I kept the breeding programme going.'

It was a bitter pill, but in a strange way the 1967 snow was also the making of Jim's success at Glenmore. During the next seven years he would slowly but surely restore the flock, and achieve some semblance of financial security.

'I'd already made up my mind about the direction in which I wanted to take Glenmore, and the big snow reinforced my feeling: we needed paddocks, we needed fences, and we needed to start growing some decent winter feed — progress. In hindsight, that time gave me the confidence to do anything. After that, the word impossible didn't exist.'

ABOVE

Loyalty.

RIGHT

Jim out mustering with the late Biggles.

Looking up the Cass Valley above Waterfall Hut, with Mt Hutton in the distance.

CHAPTER FIVE
GLENMORE — NATURAL AND BUILT

'I am forgetting myself into admiring a mountain which is of no use for sheep. A mountain here is only beautiful if it has good grass on it,' wrote Samuel Butler of Mt Cook in *A First Year in Canterbury Settlement* (1863).

ABOVE

Glennmore's homestead paddocks, with the woolshed at far left, and showing Jim's extensive shelterbelt planting.

RIGHT

The Murray siblings, from left: Gendy, Erf, Pat and Jim, October 2013.

There's a trap for outsiders who romanticise the Mackenzie Country and assume that its inhabitants see it through the same eyes. How could they? That panorama in your viewfinder is their daily workplace, and they know its rougher nature. Michèle Dominy, whose research for *Calling the Station Home* included a muster with Jim at Glenmore, cuts through the high country mystique. 'The relationship of runholders and their shepherds to their country is irreducibly adversarial,' she writes. 'They work throughout their lives both consciously and unconsciously to create a niche for themselves within the landscape . . . They survive in spite of the conditions with the expectation of moving forward, or holding ground, but never of moving backwards.'

Yet that's not the whole picture. Even now, in their eighties, having lived the thick end of their lives elsewhere, had careers, and buried loved ones elsewhere, Jim's sisters Erf and Pat still call Glenmore 'home'. This is what Jim's children mean about Glenmore 'being in the bloodstream'. It's what Dominy is referring to when she describes the pull of the high country

station 'that draws back sons and daughters to live and visit, draws back its inhabitants after trips away with a predictably steady and magnetic force'. And it's what Jim is getting at when he describes how, as he aged and farming became less of a scramble for financial survival, his response to his surroundings deepened.

Glenmore, which entered its hundredth year in the Murray family's stewardship in 2014, is 'special country', in both the farmer's and the layman's sense of that phrase. As a merino property, it has great balance. Driving up the Cass Valley, Will explains that 'for a farm to be balanced it has to have the right number of paddocks, the right amount of tussock country, of sunny country. Glenmore is set up very well. The buildings are all halfway into the property, as opposed to some farms where they're all at one end. The fact that there's a hundred hectares of paddock in the Cass also gives us the ability to do a lot more.'

That's the version you'd hear delivered at a farmers' conference. In more introspective moments, Will speaks about his favourite parts of Glenmore, the mountain country and its solitude especially, in quite a different voice.

The landscape and the heritage can't be pulled apart here. Australian sheep classer Gordie McMaster describes Glenmore as having 'charisma', much of which resides in the Murrays having owned it for so long. One hundred years as a viable farming operation is no small achievement; in these surroundings, with what has been thrown at the Murrays in the shape of both natural and man-made adversity, it's a milestone to be celebrated.

A century of farming leaves its mark on a property, even one this expansive. By the 1890s, long before the Murrays' involvement, the inventory read 40,000 acres of grassland, 13,000 acres of barren. The most recent assessment, a Boffa Miskell landscape report from 2005, cites 19,200 hectares (47,444 acres), two per cent of which is cultivated (the flats adjacent to Tekapo and in the lower Cass Valley), with a further fifteen per cent compromising lower hill slopes and rolling moraine country that has been aerially oversown and top-dressed.

Other than for fencing, the rest is as God made it, a landform for the most part steeply sloped, apart from the gentler Mt Joseph and the moraine terrace and alluvial flats. The geology is dominated by greywacke sandstones and argillite, with extensive areas of glacial tills and alluvial gravels. The soil: dry, yellow-brown earths that are shallow and stony and generally low in fertility. Because some areas of the moraines are poor-draining, there are also extensive areas of wetland and several tarns.

As for boundaries, those are set by two major braided river systems: to the west the Fork River, which flows into the Tekapo River; and to the north-east the Cass, which discharges into Lake Tekapo. A third catchment, characterised by the above-mentioned series of gem-like tarns, drains towards Lake Alexandrina.

(Anne notes that river boundaries were used for all of Tekapo's 'Gorge Runs' when they were carved out in the nineteenth century. 'Godley Peak Station is all the country between the Godley River and the Cass River, and Glenmore is all the country between the Cass River and the Fork, and between the Fork and Jollie River is Braemar, and then there's Mt Cook Station. They used to alternate the breeds of cattle between neighbouring stations — Angus, then Hereford, Angus, Hereford — so one would always know whose cattle they were.')

What else? Mountains that still wear snow in summer; steep colluvial slopes that plunge to the floors of the Cass and the Fork, creating vast U-shaped valleys; snow tussock

RIGHT

Glenmore's former head shepherd, David Grigg, mustering on Top Block.

LEFT

Stoney Tarn.

covering the high country, shifting to fescue tussock lower down, with pockets of red tussock thriving in the wetter zones. It's a simple country, with few trees and a limited palette, but by no means monolithic. Peter Newton in *High Country Journey* (1952) describes travelling deeper into Glenmore. 'At the lower end of the Cass River there are good open tussock faces and shingle tops lying well into the sun, but further up the river swings due north and the country becomes ragged and broken.' Overlooking the entire scene is the jagged silhouette of Hell's Gates.

The ubiquitous tussock gives all of this an appearance of uniformity, but it's misleading. In *Nature and Farming* (2013), Lincoln scientists David Norton and Nick Reid find great biodiversity at Glenmore, including rare ecosystems such as kettle tarns, along with the remains of pre-human woody vegetation and a number of nationally threatened plants and birds.

Between Lake Tekapo and the foot of Mt Joseph is a highly distinctive piece of country which hints at the glacial processes that formed Glenmore. Full of morainic humps and hollows known as rills, it's the home of thirty kettle tarns of recognised national significance, all of them protected under a QE II National Trust Open Space Covenant.

It is disconcerting country to walk through, says Anne. 'I remember one day Jim bringing stragglers back from Braemar, telling me to walk out and meet him, and we passed each other because there were so many humps and hollows. I got quite spooked by the vastness.'

None of which hints at the charge one feels from such surroundings, sourced partly from what you can see, partly from what you guess at, a sense of a vast alpine country at bay behind the Cass Valley. Jim and Anne's eldest, Kate Murray, now lives at Wanaka, in Central Otago, a landscape that no one could ever describe as half-hearted. Yet when she returns to Glenmore the scale shifts up a notch. 'Whether it's the vastness of the land, or the size of the mountains, or the size of the sky, or the sense of being in a basin, everything there seems more extreme.'

An extreme climate, too. 'Glenmore goes from cold to dry very quickly,' explains Anne. 'Once the ground temperature warms up, there is tremendous grass growth, second to none in New Zealand, but then it goes very dry. The autumns aren't known as good growing time, either. Some years you can get just six to eight weeks of growth time.'

Yet even with such a short, sharp growing season, this is highly productive country. 'We looked at doing forestry once and had an American company come out here to have a look around. They couldn't get over the conversion into timber, or work out why it was so huge. It's a special combination of the soils and the sun. When we were building the homestead in 1974, the builder who came up from Fairlie was very envious. He'd planted his vegetable garden at least a month before we did, but ours just bolted and overtook his.'

It's a great expanse, Glenmore, carved from monumental country, and, you suspect, a place that no one can ever entirely know. Yet over time high country farmers such as Jim develop an intimate knowledge of their surroundings. In *Calling the Station Home,*

RIGHT
Hunting sheep out — Mt Lucia.

Dominy recalls an autumn muster when her host showed on a map the 'beat' she should take (a beat being a horizontal section of mountainscape to muster).

'He supplemented his instructions as most farmers do with reference to particular landmarks such as a gate, a particular matagouri clump or rocky outcrop, that suggested that farmers know square metres of vast blocks despite the land's scale.'

By naming landmarks, too, high country farmers shrink their setting down to something on a more human scale. These names often refer to important moments from a family's time on the land, building up layers of memory and meaning, creating a sense of continuity between generations. Others are more like station in-jokes, recorded to baffle their successors.

Intriguingly, Dominy points out that the language used to talk about the high country is changing. The term 'property', for example, is now used much more frequently than 'run' or 'station', and Will Murray's generation tend to refer to themselves as 'farmers' or 'wool producers' rather than the archaic 'runholders'. The shift in terminology, she suggests, 'reflects the shift from the station of the past with a large hired staff, to the station as a family farm unit', and underlines the erosion of old social distinctions in which the early runholders were seen as a class apart.

Glenmore's names mix the poetic, the prosaic and the piss-taking in roughly equal measure, with the Devil's Staircase and Hell's Gates at one end, and Big Downs and Blow Out Creek at another (see page 100).

RIGHT

Musterers pick their way up the riverbed from Tin Shed towards Top Block. Mt Faraday lies ahead of them.

THE PAINTER'S GLENMORE

Anne Murray's watercolours of Glenmore showing, clockwise from top left: Tin Hut Creek; the Cass Valley; the road into Glenmore; the old part of the woolshed.

THE PAINTER'S GLENMORE

More of Anne Murray's watercolours of Glenmore showing, clockwise from top left: Mt Joseph; Angus and Greta with butterfly nets; Lake Tekapo; the old Glenmore homestead area.

PLACE NAMES ON GLENMORE*

AILSA PASS
This pass on the Liebig Range leads from the Upper Cass to the Murchison Valley. Named for a friend of Lady Peggy Hamilton, of Irishman Creek Station.

LAKE ALEXANDRINA
Named after Alexandrina Robinson, a sister of the Robinson brothers, who partnered John McGregor in the earliest days of Glenmore.

BILLIARD TABLE
A smooth, vertical rock and tussock face at the mouth of Tin Hut Creek in the Cass Valley.

CASS RIVER
Named for the chief surveyor of Canterbury in the mid-nineteenth century, Thomas Cass, whose legacy can also be seen in the layout of a good part of Christchurch, Timaru and other towns in the province.

DEVIL'S STAIRCASE
Sited in a basin between the Cass and the Burnett Range, this is an unusual, although possibly not uniquely, malevolent rock formation.

HELL'S GATES
A rock ledge between the Cass and the head of Waterfall Creek.

LAKE McGREGOR
Named for an owner of Glenmore.

MT JOSEPH
Named after Joseph Beswick, a pioneer of the country that eventually became known as Glenmore.

BLOW OUT CREEK
Refers to an incident in which Jack Holdom hit a rock with Gerald Murray's new tractor.

SARDINE HUT

See page 56. The setting for 'The Godley Ghost', Ernie Slow's high country ballad of a boundary keeper pursued by the Devil's daughter.

TIN HUT

Located up the Cass between Waterfall and Memorial huts and built by Gerald Murray.

THE DOWNS, INCLUDING SUNDAY, SARDINE, BIG DOWNS AND PETER'S PATCH

The last is named after fencing contractor Peter McMurtrie, who worked at Glenmore.

THE PADDOCKS

Glenmore's major paddocks are all named. 'Ray's' is named for Dr Ray Pierce, guardian of the black stilts. 'MacPherson's' is a tussock block named after a bulldozer driver. 'Clover Hill' is in honour of Harry Seivwright, who conducted clover trials in the Mackenzie during the 1950s. The 'Desert Paddock' is so-called because, says Jim, 'before it was irrigated a lizard would have had to take a cut lunch to survive'.

RIGHT
Tin Hut.

With thanks to William Vance.

ABOVE

Memorial Hut.

RIGHT

The late Ron O'Donnell, faithful musterers' cook for twenty-eight years at Glenmore, gets dinner ready at Waterfall Hut.

The strongest mark made by the Murrays on Glenmore lies in the kilometres of fencing Jim erected, in the river protection works, the irrigation and top-dressing, and the intensification of sheep numbers. The surroundings dwarf all of these efforts, and the farm buildings. Nevertheless, that built environment is an important part of station life, and each hut or shed or homestead has a story to tell.

Take Memorial Hut, for example, built in remembrance of a cousin of Gerald Murray's, Arthur Major, who was killed while mustering near the top of the Upper Cass block in 1928 — he just stepped back off a bluff in a moment of carelessness and fell to his death. Sardine Hut? 'It eventually became the skin shed, and was only the width of a skin, so was aptly named,' says Jim. Waterfall Hut, used as a mustering base in autumn, is equally self-explanatory, being located near where Waterfall Creek tumbles through a gorge in three separate waterfalls totalling a drop of 365 metres.

Built in 1916 by Mary Murray (as was the Forks Hut) of concrete, with walls plastered using the back of a spade, that particular hut is as solid as the hills behind. 'There's been a lot of whisky drunk here,' says Will, opening the door. 'A lot of bullshit talked.'

ABOVE

The family celebrates the construction of the new mustering hut.

Glenmore's engine room is its homestead, but its fine woolshed comes a close second and is a legacy of Gerald's time in charge. For his first few shearing seasons, Gerald walked his flock across a swing bridge over the Cass River to his brother Bruce's shed at Godley Peaks. But when the Kimble wool scour near Fairlie closed in 1928, Gerald quickly had it removed to Glenmore, where he and a builder converted it into a shearing shed. It was a shed built to last, made from high-quality black pine and 'Gospel Oak' corrugated iron from Australia, and is still 'as sound as a bell today', according to Jim.

By the 1980s, however, the shed was proving too small. 'We could only get three hundred sheep under cover and by that stage we were running about nine thousand sheep,' he says. 'We had no storage for the wool, so we just had to load it straight onto the truck after it was pressed. In 1981, Dick Guard and Graham Cuttle, who over the years had done an enormous amount of building on Glenmore, built a new woolshed attached to the old one — two-storeyed on the principle that sheep are shedded downstairs and carry their wool up a ramp to the board upstairs for shearing. After being shorn they're tipped down a shoot to the counting pens, while the wool is stored upstairs until there's enough to load onto the truck. Around the same time, I built new sheep yards using railway irons and pipes for the holding pens.'

Not all of the structures on Glenmore are working buildings, however. In 2012, Will built a new multi-purpose mustering hut up the Cass to use also for back-country skiing and for family weekends. 'It's about being in the mountains, enjoying the isolation,' he says. 'The kids are old enough now to spend a few nights up there during summertime and they absolutely love it.'

Getting the hut to its setting at 400 vertical metres above the valley floor was a saga. On a Saturday in May, Will's family and a group of friends tramped up to the site, while a Huey 500 flew in bunks, mattresses, material for a porch, a sink and the long drop. A larger Iroquois chopper had been organised to transport the hut from the valley floor, but just as it was about to leave it broke down. Beyond cellphone reception, the party waited for it until dusk, then walked down. All of the gear and building materials had to be left lying on site for a week until the helicopter was fixed, says Will. 'Luckily it didn't snow.'

The following weekend they tried again. This time it was the weather against them. 'We thought we had a clear window, but within about three hours we'd been hit by an absolutely screaming nor'wester. We'd only just got everything tied down when the wind whacked us, so that was a real test of the hut. It's still standing.'

HELICOPTERS

The helicopter may have played only a fleeting role in Glenmore's autumn muster, but it's been well used in other areas of station life. Jim and Anne heli-skied — they would arrange to be picked up from the homestead lawn — and Will and Ems are occasionally given a free ski trip in return for allowing commercial operators to use the top basins of the Cass.

More critically, choppers have been used to locate stock lost following major snowfalls, or to transport fencing and other equipment to the more far-flung blocks. In the 1990s, Jim began using a heli to muster deer off the tussock country.

Former Glenmore worker David Grigg was involved during both of the two autumn sheep musters when choppers were trialled. 'I remember once it was very windy, and the pilot, Gavin, told us he'd just touch the skids down for a moment and we'd have to jump out. I yelled at Robin Jamieson, who was on the other side, to make sure he let my dogs out, but he didn't hear and when the chopper headed down for a second load Jim discovered my dogs were still on board. I was left sitting up the west basin with no dogs.'

The pilot for the musters, Gavin Craig of the Helicopter Line, had a long and close association with the station. It was Gavin who made the desperate attempt to extinguish the homestead fire (see chapter 13) by scooping water from Lake Alexandrina. He was also responsible for one of the most unlikely events ever sighted at Glenmore, David Grigg recalls.

'I was working in the yards with Jim when over the hill came Gavin in his Squirrel helicopter, and swinging below it was a little car — a Fiat Bambina — wrapped up in a deer net. He'd bought it at Glentanner and thought this was the easiest way to get it home. He just plopped it down in the paddock beside the cottage, like something out of *Mr Bean*.'

The helicopter flies in to deliver the hut materials.

Waterfall Hut.

Middle Gorge, with Mt Haszard in the background.

CHAPTER SIX
JIM LEAVES PASTORAL FARMING BEHIND

The grader becomes visible from kilometres away, a burst of industrial yellow stranded among the tussock and scree of the Cass Valley. A tyre's blown, so Will is driving Jim back to the scene to repair and retrieve it.

ABOVE

Hay-making.

For a first-timer at Glenmore, the scale of the scenery in these higher reaches of the station is eye-opening. This is glacier-formed country, a geography you might say was bulldozed through by Nature. Against such a backdrop, the few man-made changes — a musterers' hut, the fence line, this rutted farm road — seem modest gestures, footprints on a mountainside.

Yet those gestures, all elements of a concerted development of the property during Jim's time, have utterly transformed Glenmore as both a productive entity and an example of high country living. Horses have been replaced by a Hilux. When the Cass floods, as it will every few years, the grader can be used to restore the road. For Jim, driving through it today, the landscape of Glenmore is full of markers of hard-won progress.

The work began in the first weeks following the '67 snow. With permission from the Lands and Survey Department, he started cultivating some land to produce hay for winter feed. Never again would Glenmore's stock be forced to scratch around for something to eat in the frozen winter hills.

The first attempt, however, wasn't exactly a roaring success. Using the tractor, Jim dragged a set of offset discs through heavy swamp ground in front of the homestead. It left a trail like a rake on bone-hard ground, he remembers.

'I sat on that tractor for a week, both hands gripping the wheel and blue

ABOVE

Mt Joseph and paddocks growing Glenmore's winter feed.

with tension for fear of being bounced from my seat or having an arm broken by the overpowered steering.'

Deciding that there had to be an easier way, he borrowed a single-furrowed swamp plough and turned over eighty acres. 'After a hundred and twenty hours of dragging that around in ever-diminishing circles I had at least broken the back of it.' That autumn and winter he fallowed it, then in spring sowed it in oats and peas, followed by permanent pasture. In the first year he made 8000 square bales of oat and pea hay, even managing a second cut of 1800 small squares. It was the beginning of the cultivation of the land to improve Glenmore.

'I realised that it was no good growing winter feed just in one area when four or five miles away there was also some very good tussock flat where we wintered sheep, so we did the same thing out at the Joseph Valley. The only hay barn on the property had collapsed under the November snow, so I quickly built a couple of replacements.'

Recognising he needed help, Jim engaged Mike Moran to drive the other tractor, a Massey Ferguson 165 that he'd bought second-hand after taking over Glenmore. Known as the 'White Chinaman' because he sold fruit and vegetables, Moran used a three-furrow to plough more land, which was again

ABOVE

Jim making 'small squares' of hay.

RIGHT

Anne.

sowed in oats and peas. The following year they cultivated another 150 acres, and top-dressed a further 1000 acres of tussock country.

'Immediately you could see a lift in wool and lambing percentages. Progressively over the next five years we cultivated more land, concentrating particularly on the Joseph,' says Jim.

Glenmore could finally produce winter feed. Given the propensity of the Mackenzie to drought, however, insurance was needed. With water readily available from the Cass River, putting in an irrigation scheme was the obvious next step. In 1972, Jim had the land surveyed for all its hoops and hollows, then engaged Twizel-based Higgins Earthmoving to form the first of Glenmore's border dyke irrigation installations.

To modern eyes, accustomed to the vast, pivoting sprinkler systems that span large tracts of Canterbury and Otago, the concept of border dyke irrigation seems quaintly low-tech. Parallel twelve-metre-wide channels are formed down a paddock, fed by water from a head-race. To irrigate pasture a temporary dam is formed, spilling water onto the land.

Border dyke irrigation has fallen from favour, seen as too wasteful of precious water resources. But in the early 1970s, none of those concerns were in play. At Glenmore, Jim didn't even have to apply for a water consent: he simply transferred the existing water right from the station's hydro plant. When that expired three years later, he got a thirty-year consent from the

Waitaki Catchment Commission to take water from the Cass for the princely sum of $50.

That same freedom to develop the station no longer applies, but in the earliest years of Jim's tenure it was across the board. Says Anne: 'If Jim wanted to put in a track or rip up a paddock or put in a fence, it was all possible — okay, so we had to get permission because we're a pastoral lease, but you could just get on with it.'

Work on the border dykes continued without pause for two months, with a new contractor from Ashburton tackling the earthworks using two graders, a brace of scrapers and two men on the ground. Using borrowed moulds and a concrete mixer brought up from Fairlie, Jim and two men built the dam structures. Tough physical work, it took them five weeks to finish, but for Jim it was an exhilarating time. His Glenmore was taking shape.

'To see the production coming off such incredibly light lands was impressive,' he says. 'Prior to growing our own feed, our normal winter losses would be up to ten per cent for ewes wintered up the Cass. To be able to feed them during the winter was a hell of a breakthrough. Coupled with the increased wool weights, lambing percentages and general stock well-being we were making wonderful progress.'

By now Anne was very much part of that equation. Given how seldom Jim left Glenmore in the early years, it was a stroke of luck that they met at all.

JIM LEAVES PASTORAL FARMING BEHIND

In April 1970 Jim had skipped the final day of mustering to attend a friend's wedding at Darfield. There he met Anne Le Cren, a landscape architecture student from Lincoln with ties to the high country — her father was born and raised on a station on the other side of Tekapo from Glenmore. They were engaged by July and married the following January.

'That was how it was in those days,' says Anne. 'If you fell in love with somebody and you'd decided to marry them, then you just got on with it.'

An extended honeymoon was never on the cards, in any case. Jim was busy. When the first phase of irrigation proved so successful, Jim decided to add another seventy-three hectares, quadrupling the original coverage. Cultivation, too, continued apace. And with fresh paddocks to protect against the drying nor'wester, shelterbelts became a priority.

Planting trees in the Mackenzie can be a heartbreaker. 'The combination of poor soil types, severe winds, hot summers and cold winters means it's a constant battle to get them established,' says Jim.

They concentrated on ponderosa and Corsican pines, planting 1000 to 1500 each year, losing perhaps forty per cent to drought, frosts and pests. Hundreds of trees were ringbarked each winter by rabbits and hares foraging for food when Glenmore was under snow.

Fencing went up — mile upon mile of it, breaking massive single blocks into smaller units. Where a tractor could access the high country, posts were erected every nine feet; where that was impossible, they settled for two posts to the chain.

RIGHT

Top Block.

The Cass Valley, which at the time was divided into four summer blocks and one wintertime block, proved particularly tricky. 'The problem was that when we turned the sheep out in the summertime they all ended up back on the winter country. We needed some kind of permanent boundary in the riverbed at the end of each fence line. So we drove railway irons out into the riverbed in a line, hung a wire rope through them, and tied netting to the bottom side.'

A combination of relentlessly hard work and canny stewardship had begun to tell. Glenmore's overdraft was still in the red, but the bank manager could see that Jim and Anne were making progress on the path they'd set and was sympathetic to the cause. For the most part, Jim held firm to his commitment to spend only what the station earned the previous year.

'We farmed at a wonderful time, when new farming technology and techniques were coming into the Mackenzie,' says Anne. 'The limiting factor of farming in the high country was always being able to winter your stock, so by making hay and growing feed you immediately lifted your potential. People were beginning to have access to the canals for water, and irrigation was creeping in to other dryland parts down-country. So our timing was very good.'

Financing was also freely available. Through the late 1970s the Muldoon government introduced a series of farm subsidies, meant to offset the double whammy of a high exchange rate and expensive material imports while boosting primary exports. The Land Development Encouragement Loan made cheap loans available to develop unproductive land, while the Livestock Incentive Scheme encouraged farmers to carry more stock. A third leg, the Supplementary Minimum Price scheme, guaranteed farmers price stability even as export values fell.

Looking back now, Jim describes the thinking as 'suicidal' economics, but at the time the subsidies were an offer too good to refuse. 'You grabbed it because it was there,' he remarks.

Using a cheap loan, Jim cultivated an additional 120 hectares on the Mailbox flat, employing Brian Groom from Fairlie to tackle the work. The first year he grew oats, but soon discovered the limits of his experience.

'I bought a grain header, which was one of the most stupid things I ever did. I had absolutely no knowledge of how a header worked, but I was determined to get a cash-flow return from these oats. I can still remember wobbling around on that machine, with the occasional fire breaking out in the engine — obviously the result of not cleaning it thoroughly enough. In wetter areas we had to tow the thing behind the tractor. But in the end we got three hundred tons off that first cut, and I put up lots of temporary silos and

RIGHT

Will and Ems bringing in a mob in late April, with an early fall of snow on the hills.

LEFT

Ewes in the yards near Glenmore's woolshed at weaning time.

sold quite a lot of grain to recoup some of our costs.'

Aided by a programme of aerial oversowing and top-dressing, Glenmore was now producing a good amount of winter feed. Jim was in a position to add more stock. Again, his timing was propitious. Like every pastoral lease in the high country, Glenmore had a stock limitation imposed by the Department of Lands and Survey, assessed and regulated by the Chief Pastoral Lands Officer. 'We had to apply to increase the numbers, but they were very encouraging of it. Approval was just a formality.'

While all this productive development was underway, Jim was keen to open up the backblocks to make Glenmore easier to farm. With the day of the helicopter farmer some way off, this involved building tracks through the high country, taking advantage of a forty per cent subsidy offered by the Catchment Board to encourage firebreaks. While the tracks also served that purpose, they were mostly used to bring in fencing supplies and as access for mustering.

About the same time, Glenmore's neighbour Godley Peaks was beginning to retire some of its high country, in return for funding to develop land closer to the lake. It was called a run plan, and Jim didn't want a bar of it.

'There was pressure coming on from the Waitaki Catchment Commission. It was all aimed at the problem of soil erosion — got to close this land up, remove any stock, let it regenerate. There was a financial carrot dangled with a grant to put into improving other lands. But it was bullshit and it went completely against my grain, because it

ABOVE

The muster is over and the sheep are in: Andrew Steven, Jim and David Grigg returning home.

would have upset the balance of the property. Glenmore is a beautifully balanced property. You retire country up the top to improve the lower country, you're giving up your ability to give the lower country a bit of a spell at times to regenerate.'

Taming the rivers was the final piece in the puzzle. With Glenmore's river flats being eroded at the rate of half a chain a year, something had to be done to keep the Cass River in particular to its natural course. The answer: a brace of river protection works, including a mile-long system above the homestead.

Built on a system of piles, wire ropes and diamond netting, with a stopbank in behind for immediate control, the new infrastructure proved expensive but effective. Jim added long-term protection by planting 17,000-plus poplars and willows.

'All the manual work we did ourselves,' he remarks. 'If you want to know how laborious it was, I once worked out that we had put on 15,000 ties joining netting to rope on the works above the homestead.'

By the 1980s Glenmore was transformed. The rump of a flock left by the 1967 storm had grown to 12,500 sheep, with 400 breeding cows. Wool weights and lambing percentages were up, and surplus stock was fetching decent returns. 'We'd gone from nothing to making thirty thousand small

squares of hay and then to making three thousand larger round bales, and then to three thousand tonnes of silage plus five hundred rounders,' adds Jim. 'Our costs were up, but productivity had climbed, and more importantly the returns on that productivity were way up there. We were now on a constant development roll.'

But 1967 wasn't forgotten. Jim knew that good fortune had no immunity to bad — another storm, an outbreak of disease, a tumble in the price of wool, and Glenmore could easily be at risk again. The next step was obvious: he had to diversify.

— ✕ —

During Anne's first winter at Glenmore, a pipeful of water froze under the ground. It was 1971, the days of the old homestead, and household water still had to be pumped every day, a ritual that involved shutting six taps outside and pumping away, then reopening the taps to drain the pipes. Anne had made the beginner's mistake of forgetting one of those taps.

'It took the men a whole day to axe up the frozen ground and then work their way along the pipe with an acetylene torch to thaw the frozen water.'

It was a rough introduction to the high country winter, but nothing she couldn't handle. Born Catherine Anne Le Cren on 7 June 1945, in Gloucestershire, Anne came to New Zealand by ship aged two, with her brothers Mike and Chris. Their parents Philip (Pip) and Marcia had sailed six months earlier on a troopship, intent on finding a farm in New Zealand. After camping out for several months, the Le Crens finally found the property they wanted in North Canterbury's Leader Valley.

For Anne's mother Marcia, raising a family on a New Zealand farm was a world away from her own upbringing. The daughter of an eminent orthopaedic surgeon, she had enjoyed a privileged childhood in London. Pip, however, was in far more familiar territory: his family had farmed the shores of Lake Tekapo, enduring the rigours of huge snows and stock losses. His was not a privileged background, but thanks to an uncle's generosity he'd been able to attend Cambridge University, where he read science. What the couple had in common was a vigorous work ethic, vital for anyone wanting to farm in Canterbury.

BELOW

The Le Cren family: Anne, Pip, Marcia, Mike and Chris, 1961.

Anne contrasts her childhood strongly with Jim's.

'My parents were very young, very liberal thinking and broad-minded, and we did all sorts of crazy things together — camped and tramped, skied, had huge adventures as a family. All of that has been new for Jim, and I've always admired how he's run with it.'

The differences extend to communication, she adds. 'Jim's family weren't demonstrative in any way. They were reserved and didn't communicate much beyond practical matters. And again, as a family, Jim and I have done an enormous amount of talking with all our children.'

After Parnassus Primary, Anne boarded at Christchurch's Selwyn House, where the syllabus was set in England. 'It set the precedent for me having a wide range of interests for the rest of my life, particularly art, geology and the environment,' remarks Anne, who completed her schooling at Craighead Diocesan School for Girls in Timaru.

After studying for two years towards a BA, she left for overseas, hitching and camping her way around Europe and reconnecting with family in the UK. 'I learned far more travelling than I ever did at university,' she remarks.

She also belatedly recognised where her passions lay. Following a brief stint with the Department of External Affairs' consulate division in Wellington, Anne enrolled at Lincoln to study landscape architecture. Which is when Jim entered the picture, at the wedding of a childhood friend.

'Jim told me that he'd lost a day's mustering to come to this wedding and I replied, "Well, look at the benefits." We clicked straight away. A few weeks later he asked me, in a roundabout way, to marry him.'

The proposal was straight out of the drawer of taciturn rural Kiwi male. 'How would you like to come and live at Glenmore?' asked Jim, before adding: 'You could feed the musterers and shearers for me.'

Anne completed the year at Lincoln, but there was no way she was going to finish the course. If she hadn't moved to Glenmore, she says, she'd have barely seen Jim. 'He had full responsibility for the place with a large debt around his neck, so his focus was very much on financial survival.'

Her first year at Glenmore was more difficult than she had imagined. 'My childhood had been on a farm and my mother was a very practical person. Both my parents just got on with things, and I thought I'd marry Jim and do the same. But that first morning, Jim, the single boy and the married man all headed off, with Jim saying, "We'll be in for morning tea", and I thought, Uh-huh, is this to be the daily routine? I was sometimes lonely that year.'

Work helped. Anne was busier than she'd ever been, cooking three meals a day not only for Jim and his two staff, but also for anyone else who was at

LEFT
—
Ewes coming down onto the Old Glenmore Block by the lake.

Glenmore. They included as many as fifteen shearers for three weeks of every year, teams of crutchers, hay carters in the summer, musterers, fencers, machinery men, silo contractors and stock agents. When the border dykes went in there were eleven extras to feed along with the shearers, and Anne had to cater for two sittings.

'Whether they were coming to stay or simply arriving for a short visit to do with farm business, they expected to be fed. It was a high country tradition.'

Within four months of marrying, Anne was pregnant. Katherine was born in January 1972, followed by William in November that same year, and Phillipa in April 1974. 'We had three children within two-and-a-half years, which I don't recommend. I was permanently pregnant or with babies.'

She needed help. Cue the arrival of Catherine 'Tid' Alston.

An Australian whom Jim and Anne met through family connections, Tid was the proverbial lifesaver. 'She was from a farming family just out of Melbourne and was the most wonderful person. She was thirty, unmarried, very musical and the warmest, most intelligent, fun-loving person imaginable. She committed to stay for six months and became a very firm friend, keeping the show on the road for us. Well, that is apart from when the Melbourne Cup was on! I can remember Tid getting wildly excited, and babies and everything else had to wait while she listened to the racing.

'She could cook and care for children, and was sensible and practical. She could also manage the vege garden, handle the telephone and drive to Timaru on chores.

LEFT

The Glenmore yards during weaning.

I would have gone under without her there, I think.'

Soon after Tid left, Anne was pregnant again. 'We engaged Kay Findlay, another wonderful Australian girl. She ran the domestic side of things and stayed for a long time after Pip was born. We also commissioned a Karitane nurse with the intention that she would help Kay while I was in hospital having Pip, and were given a German woman. She was very brusque, and I remember having this sinking feeling: How nice are you going to be to my two babies? She drank whisky at night, knew where her duties began and finished, and told me exactly how hopeless a mother I was.'

The new help didn't last long. One day while Jim was visiting Anne at the hospital, Rosalie Mason, wife of the head shepherd, saw the nurse heading off in the Toyota to visit the single men at Godley Peak. Checking the homestead, Rosalie found Will and Kate alone, having their afternoon naps.

Notwithstanding the occasional home-help drama, life at Glenmore was full and happy. Anne says she has always felt a 'huge sense of privilege' to be raising a family in such a place. The high country social life, too, was rewardingly rich.

The annual Mackenzie Federated Farmers Social Club Ball was a highlight of the year, attended by most of the runholders, their wives and girlfriends, and the people who ran the rural service businesses. Anne notes sadly that the ball, along with other traditional farming events, has vanished as a result of farmers' lives becoming busier and more complex. The big Tekapo family picnic disappeared when sales day was shifted to accommodate cartage company demands.

The Murrays often entertained at home. 'With the slower pace of life we tended to make our own fun, and there were frequent dinner parties. They were good, old-fashioned, hilarious parties. There was no age barrier, because there were only twenty or so properties in the basin and so you invited everybody. And no matter the age, we all called each other by our Christian names — it took me a long time to get used to this informality.

'Peggy Hamilton, the widow of Bill Hamilton, the jetboat inventor, was well into her eighties, and she was fantastic fun. If there were stories going around the table she would call a halt until she'd changed the battery in her hearing aid so she didn't miss anything. Bruce Hayman had been a bomber pilot in the war who'd crashed on Mt Etna. He was a fascinating raconteur, and so his stories were definitely paused until Peggy's batteries had been replaced.'

Jim and Anne's closest friends were neighbouring farmers Gill and Hugh Hunter-Weston of Mt John Station, and Carol and Duncan Mackenzie of

ABOVE

Anne and Jim.

Braemar. 'With five children between them, they became like an extended family for us, and we shared much of our life in the Mackenzie as well as the children's Correspondence School lives.'

Among these young married farming couples, there developed a tradition of elaborate practical jokes. Anne recalls the time she and Jim took advantage of their neighbours' absence on holiday to put an ad in the paper on their behalf requesting photographic models — they were inundated with messages. On another occasion she 'tarted up' a piece of New Zealand Defence Force equipment.

'An army quad truck had broken down between Balmoral Military Camp and Tekapo and been left on the side of the road. I went out and painted huge flowers over it. Nobody knew who'd done it — or so I thought. I was at a dinner party and heard Michael Murray, who was then chairman of the Mackenzie District Council, speaking to someone behind me about this act of "vandalism", saying he'd heard that the Ministry of Defence was furious and were on a witch-hunt to find the culprit. Turns out he was just having me on, but I fell for it.'

Anne and Jim always appreciated that the success of Glenmore depended on others. In recognition, every Christmas they invited anyone who'd helped during the year to a sit-down dinner.

'One year, however, I told our single boys, tongue firmly in my cheek,

"Stuff it, I feed you all year, you can do the dinner this year!" They rose to the occasion and formally invited us over to their quarters for Christmas dinner. We were greeted by all these fellows, all dressed in white shirts and black trousers, with saucepans on their heads. They served us a Christmas dinner "off the land", starting with trout from the creek at the bottom of the garden, followed by rabbit stew and a trifle pudding. We had the most amazing evening.'

As the farming environment grew more complicated, Anne became increasingly involved in the 'bureaucratic hassles' Jim faced. As well as helping with routine admin chores, she spent months dealing with district plan changes and helping with submissions and other paperwork. 'I was a mother, and I was a station cook, but I was also very much in partnership with Jim when it came to running the farm,' she says.

As if life weren't busy enough, in late 1974 they took on building a new house. 'We were moving at such a fast pace in those days. Jim was moving flat out on the farm and I was having babies galore, but you just got on and did things; you just worked.'

And change was needed. Glenmore homestead had been built by Mary Murray in the early 1900s as a four-bedroom manager's house, the third dwelling on Glenmore. The original homestead had been situated at the Cass flats — you can still see the remains of the foundations among the paddocks. The second, a cob cottage on the shores of Tekapo, was demolished in 1952 when the lake was raised.

Like all Mackenzie homesteads, Mary's had been built near the water and out of the wind, meaning it was low to the ground, with any outlook blocked by shelter trees. When Anne arrived at Glenmore she lobbied Jim to take out the worst offenders, exposing a view of the mountains at the head of the lake.

Yet the kitchen in particular remained a bleak space, positioned on the south-west of the house with little view and less sun. It was so frigid, in fact, that Anne looked forward to winter when the coal range would at least be burning. A large storeroom on the south side often got so cold it became difficult to keep preserves from freezing.

The house's only bathroom encapsulated the homestead's shortcomings, with floors on a tilt so that if you left a pram at one end you could collect it after your ablutions at the other. Lying in her bath at night, Anne watched spiders crawling from holes in the walls.

All of which could be endured, possibly. But there was a hazard to health here: Pip, the youngest, was hospitalised for pneumonia after sleeping as a newborn in a bedroom off the kitchen.

Calculating that it would be cheaper simply to build a new house, they chose a warmer and drier site up the hill and drew up plans for a six-bedroom dwelling, surrounded by a verandah to keep the weather away. They designed it themselves to save money.

It took Fairlie-based builder Alan Johnson sixteen months to build the new homestead, helped by Jim and his staff. They made concrete using shingle from the Cass riverbed, and the cladding from greywacke boulders that Jim and Anne hand-picked at Godley Peaks. The handsome Oregon timber was milled from trees that Gerald had planted at Godley Peaks.

'All I can remember is shovelling the shingle,' says Jim. 'It was five thousand square feet, using two concrete mixers, and I just shovelled and shovelled and shovelled.'

Nothing was wasted of the old homestead. The verandah and storeroom were added to the single men's cottage, the band rotunda-like master bedroom was deployed as a day hut up the Joseph, while the bulk of the homestead was trucked by a Brightlings transporter to the shearers' quarters, where it is used today as guest accommodation. It was a home makeover high country style, and not without incident. When the communal bathroom tipped off the transporter and crashed through the septic tank, remarks Jim, 'we were in the shit properly'.

Moving into the new house was a relief. It faced the sun, offering a view of Mistake Hill and the lakehead mountains. The walls and ceilings were well insulated, there was double-glazing, underfloor heating, a double garage and an easily accessible woodshed. 'I remember the warmth of the house,' says Anne, 'and the sense of space, and the feeling that it was all hugely smart.'

Better yet, the water no longer had to be pumped.

From halfway up Glenmore's Big Downs the view shifts from merely arresting to breathtaking. Immediately below is the turquoise brightness of Lake Tekapo; to the right, Lake Alexandrina, its surface darker, wind-ruffled. This high front country is part of Glenmore's deer operation, perhaps the most successful of Jim Murray's early punts.

The idea that deer could be farmed had been no easy sell in the late 1960s — at least, not to those for whom the official designation of deer as a 'noxious animal' said everything. Introduced for sport in the nineteenth century, deer encountered an ecosystem that might have been designed for their convenience, with no predators and as much food as they could ever need. Populations grew rapidly, and by the early twentieth century farmers and foresters had a new pest to hate. A 'Deer Menace Conference' was held in Christchurch in 1930, following which open season was declared.

Hunters made hay. By 1956, when the Noxious Animals Act shifted responsibility for deer control to the Forest Service, government cullers were killing 50,000 a year. And when deer meat began to fetch good prices and helicopters became available for recovery, those numbers climbed exponentially. On the eve of deer farming becoming legal in 1969, the best hunting operations were tallying 200 carcasses a day.

The first deer-farming licence was taken up by a property at Taupo. For the times, it was a relatively tightly regulated farming activity, and would-be deer farmers needed approval from three separate authorities: the Ministry of Agriculture and Fisheries, the Forest Service and the Ministry of Works. But compared with today, says Jim, it was mercifully free of other red tape.

From small beginnings, an internationally dominant industry would be born. By the start of the 1980s there were 1540 deer farms, and by the end of that decade New Zealand accounted for half of the world's farmed deer. In the following fifteen years, the number of specialist deer-farming operations grew by fifty per cent, a significant new phase for what had begun as a sideline operation for most farmers.

Back in 1970, when Jim decided he needed to diversify his farming, he had no sense that deer would provide the windfall to finally clear Glenmore's debt. Deer farming had only been legalised the previous year, and in the Mackenzie Country no one had shown much interest other than Haldon Station's enterprising James Innes. Jim, however, saw straight away that deer could work on Glenmore — in fact, the back country was already rife with the things. 'Deer love this environment, and their fawning time in late November and weaning time in early May suited perfectly.'

As for the money side, it was too early to tell how the new venture might go, but demand was growing strongly for New Zealand venison and there seemed to be a market emerging for red deer velvet.

Glenmore became the seventeenth New Zealand farm to be issued a licence to 'keep a noxious animal in captivity' — officialese of the day for farming deer. 'It gave us another string to our bow,' says Jim, who reveals that prior to taking on deer they had seriously investigated growing peonies for export, only to be stymied by the complications of marketing overseas. 'With deer, we were no longer solely reliant on wool, surplus stock and cattle. It eventually became a very good secondary income stream.'

He spent eighteen months researching the new farming techniques and

LEFT
Hinds race away from the deer yards after testing.

putting in deer-proof fencing and a deer yard, using former telephone poles from the dismantled Pukaki and Mt Cook lines, and car packing cases of Malayan hardwood for the walls.

The swampy flatland below the homestead was chosen for the bulk of the herd, with the higher downs country reserved for fawning and weaning. 'We'd had cattle on the flat, but I believed they were too hard on that land, with all their pugging, and it was very cold in wintertime. So it was a good fit for deer.'

When it came time to stock the enclosures, however, Jim was dismayed to find that helicopter-borne 'poachers' had got in first. 'I used to go out mustering and see mobs of sixty or seventy deer. Obviously people had realised there were quite a few deer on Glenmore and helped themselves. There were no rules in those days.'

(Footage from the early days of live helicopter deer recovery is of the type you tend to watch between your fingers, mouth agape. Young men leap from the air onto fleeing stags, 'bulldogging' their prey into submission. A chopper clips a tree and survives, another becomes wreckage. It's an OSH inspector's nightmare brought to life in the high country.)

Jim was forced to buy in animals to supplement what he could find on the property, a total of forty hinds to kick-start Glenmore's new venture, all of them red deer. 'I always tried to ensure they were Rakaia reds, which were considered a superior bloodline. Later you started seeing wapiti being crossed with red hinds to make the progeny grow bigger, and then the arrival of

European reds, but our breeding programme was always about retaining those initial Rakaia red bloodlines. We just stuck to what we knew.'

Jim remembers his early experiences with deer as a steep learning curve. 'There was great communication between all of us farming deer in those days. We talked among ourselves: "How do you handle these damn things?" There were no designed deer-handling facilities, although people had all sorts of theories. It soon became apparent that if you had a roof on and made the yard as dark as you could, then you could handle them more easily.'

The key was patience, a softly-softly approach that didn't stress the animals. 'Deer are the most wonderful animals to work with, just incredibly responsive. If you walk into the yard thinking you can handle them quickly, well forget it. You need to talk to them before you open the door, let them know you're coming in, and then handle them very gently, talking to them all the time. They respond accordingly.

'But at the start we humans were possibly as apprehensive about them as they were about us. I remember, for example, the first time we brought in stags for velvet. We pussyfooted around, and eventually got them into the yards. At the time, the favoured method for retrieving the velvet was to dart them with a tranquilliser dart fired from a blank .22 rifle, a method developed by Colin Murdoch in Timaru. Well, we didn't get into the pen with them; instead we made holes in the wall of the pen and shot the darts into their rear ends, all of it supervised by the local vet. Those first two or three years were

RIGHT

Some of Glenmore's deer herd at weaning time.

exciting, but there was also a degree of fear because we had little experience in handling such big animals, especially the stags.'

Initially, velvet was the sole source of income, with venison added later. 'We ploughed any cash flow from velvet straight back into the deer. From the start, the operation had to stand on its own two feet, with no cross-funding from other parts of the farm. It had to be that way, because we didn't know if this was going to last. We didn't know if velvet prices would hold up, and the same with venison prices. Should we hang on or walk away? It was a risky undertaking.'

The gamble paid off. By the mid-1980s, hinds were fetching $3000–$3500 apiece, an unbelievable price. Having built up Glenmore's herd to 280-odd hinds, Jim was tempted to sell large and dump the farm's debt. They had borrowed more on top of their mortgage — 'against my better judgement, I must say, because I'd always maintained that one should develop a property out of cash made the year before'.

But this was the deer industry's moment in the sun; the market was humming and apparently only better times lay ahead. Jim was strongly advised to hold his hand, to keep spending and developing. 'Throughout my farming career I've always picked the brains of people whose expertise I respected, and the advice I was getting was "Jim, don't get rid of the mortgage, just carry on."'

They went against the advice. 'I've always maintained that in your farming career you'll get a financial spike about once every decade to twelve years, and that the key is to make the most of that spike, be it paying off the mortgage or, if you've only got a small debt loading, then investing into something off-property. We sold the deer, wiped the mortgage, wiped all our debts, and it was bloody fantastic, we never looked back.'

Having dumped the mortgage, he spent the next years building up deer numbers again, until Glenmore's herd totalled around 300 hinds and 200 stags. 'It became a very good little enterprise.'

But what was that line of Robert Pinney's? 'No high country man is ever complacent.' In a single afternoon in 1987, Glenmore's wonderful run with deer turned into every farmer's nightmare.

'I had always been incredibly careful to ensure that the animals we brought in had had two clear TB tests,' says Jim. 'It appears this wasn't sufficient. During some routine testing, we discovered we had the disease in the herd.'

Nine deer were taken up to the killing shed for confirmation by a vet. When they were opened up, the sight was 'hideous', according to David Grigg, head shepherd at Glenmore at the time. Everyone was shell-shocked.

ABOVE
Glenmore's stags.

'I had all the deer yard keys in my pocket and I went down to the irrigation paddocks,' he recalls. 'Suddenly this old Holden came flying down behind me — it was Jim, yelling "Where are those bloody deer keys!" He was very distraught. He locked the sheds and called everyone back to the homestead, where we all had a few stiff drinks.'

Several top investigators from the Ministry of Agriculture and Forestry soon arrived from Wellington. Glenmore was among the first Mackenzie Country properties to be hit by tuberculosis among deer, and there was an urgency to contain the outbreak if possible. The infected animals were killed and buried. But as more TB was identified at other farms in the Mackenzie and elsewhere, the ministry instituted stock movement controls. 'We couldn't shift any stock out of the Mackenzie for sale other than to the works. It was strict.'

The investigators mistakenly assumed that the TB must have come from possums, not understanding that the explosion in rabbit numbers in the Mackenzie had also attracted vast numbers of ferrets, a highly mobile pest capable of covering twenty-five kilometres in a day. Once it was confirmed that ferrets were the vector, the solution to the outbreak was obvious: get rid of the rabbits, and the ferrets and TB would hopefully disappear. It worked. 'We did some rabbit control work at Glenmore and the TB incidence dropped right down,' says Jim.

In the meantime, however, the TB outbreak put a crimp in Glenmore's nascent deer operation. It took three years before Jim was able to get it back on track, during which time he invested nothing more in deer.

The Church of the Good Shepherd.

CHAPTER SEVEN
THE NEIGHBOURS

Looking west from Mt Hay Station, the eye is drawn magnetically across the lake to Glenmore and its backyard alps. Anne's late father Pip grew up on this side of Lake Tekapo when part of Mt Hay was known as Tekapo Station, before the lake was raised and the old homestead drowned. He grew to love Glenmore, too, after Anne married Jim, but he always maintained that the view of it was better than the view from it.

On this morning, the paddock between the Mt Hay woolshed and the lake is heavily parked with four-wheel-drives. Mt Hay is the third stop of the Mackenzie District Merino Stud Tour, and dozens of local farmers and others have dropped in for an hour. If there was ever a gathering of the Mackenzie clans this is it — a day for runholders to get together over a cuppa or a whisky and compare notes.

Gordie McMaster, an Australian sheep classer whose influence pervades the merino runs of the Mackenzie, reckons it's the harsh environment that makes this community work.

'It's just like in Australia: the further out you get, the closer the community,' he says. 'You cross Burkes Pass and you see space and big skies, and only then you start to spot the stations. It's not like further up in Canterbury where there are lots of houses — here it's mostly all space. And it's just so bloody harsh. These families, a lot of them, have been connected for a long time. And they've all been through hard times.'

At Mt Hay, the runholder's wife Sue Simpson has put together a small home museum in a 1911 musterers' shed: exhibits of farm journals and diaries, photographs, and displays of vintage farming equipment. Mostly it's about Mt Hay's history, but as the visitors file through there's the odd exclamation of recognition at an ancestor's face spied in a photo, or a journal reference to another forebear. A discussion begins about historic shifts in the ownership of particular Mackenzie merino runs, the early cross-pollination of the basin as families bought and sold and traded and partnered up.

At their lowest moments, the Murrays have been able to draw on the very best of this Mackenzie community spirit. After snowstorms and house fires, people arrived at Glenmore in droves to help. But the family has also given back — played a leadership role, in fact.

'There has been this sense of responsibility towards the wider Mackenzie Country community from the Murrays, and it started with George's generation,' says Anne.

The famous Church of the Good Shepherd at Tekapo, for example, was built on land that was donated in 1933.

Months earlier, George and his sons Gerald and Bruce had met with the Rev. Walter Davies to hear him outline his plans for a lakeside church to serve runholders. It would be built to the glory of God, but also stand as a memorial to the pioneering runholders and shepherds of the district. A committee was formed, and a site near the old Tekapo bridge scouted as a fitting location.

It was Dick Beauchamp, Bruce and Gerald's brother-in-law, who saw the land advertised for sale in the *Timaru Herald*. 'We realised that it included the same site that we wanted for the church,' Dick once recalled. 'They were pretty hard times and we had no cash to spare, so I rang George Murray and asked him to buy the land.'

Wearing his best poker face and that of a lemon at the auction, George bought all four lots, outbidding several disgusted fishermen wanting it for lakeside baches. When he donated the land to the church, he stipulated that it must be left in its native state, right down to the matagouri. As Jim observes, he couldn't possibly have anticipated the huge increase in foot traffic.

In later years, a well-supported group of church caregivers was set up, with families such as the Gould Hunter-Westons to the fore. They had their work cut out dealing with the impact of tourism. Even in the early days of Jim's eighteen-year stint on the vestry, the church was getting thirty-five busloads a day during peak season, plus independent travellers. Wear and tear and vandalism were taking a toll, and land around the church was worn down to bedrock.

'We'd been relying on ad-hoc maintenance, but no one was really looking after it, so I set up a committee of six people in Tekapo to make it happen. We put in asphalt footpaths and an irrigation system, and edged the front paths in stone. As much as possible we adhered to George Murray's wishes and only used grass, tussocks and matagouri planted around the footpaths.

'We also employed several guides to mind the people. The money for some of their salaries and for the maintenance of the church and grounds comes out of the two donation boxes. Unlike some churches in New Zealand, thankfully we're at least able to break even.'

Another Murray endeavour was the Roundhill ski field. Jim was one of four young runholders who set up the field at Tekapo, working with Bruce Scott of Godley Peaks, David Allen of Bendrose and Brian Waters of Richmond Station.

In the early 1960s, creating a new ski field was a cinch, according to Jim. 'As the lessees, Richmond Station had to write to Lands and Survey asking if they were happy for us to put a ski field on pastoral lease — the tick came back straight away. Bruce Scott took his TD15 dozer, Richmond had a D6, and before you knew it we had punched in a track on the sunny side of the terrace. Brian Waters' father Don found an International truck that had been used for towing gliders, and that was taken up there, an anchoring weight with a rope pulley was sunk on top of Roundhill, and two big sledges were

THE CHURCH OF THE GOOD SHEPHERD

It's become a small high country tradition in its own right: visitors to the Church of the Good Shepherd will conclude by filing down the path to build a cairn at the lake's edge. There are hundreds of these things, little stone offerings to the gods of tourism, teetering before a million-dollar view.

Possibly it's all inspired by the church's memorialising spirit. The building was designed by Christchurch architect Richard Harman to plans drawn up by Mrs Norman Hope, of the Grampians Station, and the very robustness of its walls and fixtures is a testimonial to the toughness of the Mackenzie pioneers. Lakeside boulders, left unsculpted and still with a covering of lichen, formed the walls, with red pine used inside. Again working from drawings by Mrs Hope, sculptor Frederick Gurnsey created a stone altar, altar cross and an oak lectern stand that are solid and simple rather than sentimental. In keeping with the theme, Gurnsey's carving of the Good Shepherd is also more robust than the usual representations of Christ.

'Those who survived these early times knew what it was like to live on a knife edge of existence,' wrote the Rev. Peter Hurricks, a former vicar at Tekapo, in a booklet for the church's fiftieth jubilee in 1985. 'The Church of the Good Shepherd was an attempt to express this pioneer experience in a living church of stone.'

People from around the Mackenzie were encouraged to donate fittings in memory of specific pioneering family members. The church bell, for example, was gifted by the Barker and Sealy families to honour Audrey Barker and her grandfather Edward Sealy, alpine explorer and surveyor. The Book of Remembrance was given by the family of Thomas Teschemaker of Haldon Station. And so on, right down to the bronze candlesticks, given by the station hands at the Grampians. Elsewhere, broad groups of pioneers are honoured. The font, for example, is a monument to the station hands and shepherds of the Mackenzie, while the vicar's chair is a memorial to the district's pioneer women.

The church was officially opened in January 1935, with the Duke of Gloucester laying the foundation stone after a service conducted by the Bishop of Christchurch. While officially part of the Anglican parish of Fairlie, it is administered by the Anglican Church Property Trustees in Christchurch. The church was never consecrated, leaving it open to all denominations.

Looking up Lake Tekapo from inside the Church of the Good Shepherd.

added. Someone would sit on the International, changing gears all day, while the sledges carted people up the slope. It worked brilliantly.'

When it became obvious that the new field was a hit, a rope tow run with an Armstrong Sidley air-cooled motor was ordered from Andrews and Beaven, an engineering firm in Christchurch. Jim volunteered to collect it, accompanied by Lars Anderson, the gardener at Godley Peaks, a Second World War veteran with a history of drinking benders. 'I was loading the motor and tow onto the back of the flat-deck Volkswagen truck when old Lars just disappeared. I spent the next three hours wandering around Moorhouse Avenue, going in and out of pubs looking for him.'

After Richmond Station changed hands, Roundhill was run from 1974 as a commercial field by Karl and Audrey Burtscher, whose family have since added snowmaking and the world's longest rope tow. It's a very different beast now, but Jim remembers those early, seat-of-the-pants days with affection. 'We used to have these big open days up Roundhill, all the locals came up, and they all brought food. We'd have a good day's skiing, then a barbecue, and then leave at seven or eight o'clock at night, in the dark. It was great fun.'

Jim was also heavily involved in farming and wool-related organisations, serving as secretary of the Mackenzie Branch of Federated Farmers, with ten years as a member of the South Island High Country Committee ('it had huge political clout'), several years on the inaugural board of Merino New Zealand, and several more on the Waitaki Catchment Commission.

His longest-running commitment, however, was as a farming representative on the Tekapo Pest Destruction Board. As will be seen in a later chapter, he still laments the end of these highly localised bodies.

'As pastoral lessees we were totally responsible for the control of our pests. They were very efficient organisations. We planned twelve months in advance to ensure runholders could make appropriate stock adjustments for poisoning

BELOW

Scenes from the early days of the Roundhill ski field in the early 1960s.

operations and to give the carrot growers around Temuka plenty of notice.'

During his twenty-seven years on the board, he watched aerial pest control operations in particular advance significantly, from bait being carted in sacks and deployed by 'someone sitting in the back of an Auster aircraft feeding carrots down a chute in its belly', to bulk loaders and mechanical release and turbine aircraft.

The state of the merino industry was another concern. Gordie McMaster remarks that Jim has always wanted the Mackenzie to be seen as New Zealand's top merino district. 'Jim was without doubt a leader there,' he says. But Jim also took the attitude that a rising tide lifts all boats. Knowledge should be shared. Fifteen years ago, Jim established the Merino Educational Trust, aimed at helping young people in the merino scene here spend some weeks working in the industry overseas. Initial funding was sourced from the runholders of the Mackenzie/Waitaki area, and the recipients were expected to use whatever they learnt overseas to help improve the New Zealand scene.

'The past couple of years we've had people go to South Africa as opposed to Australia,' says Jim. 'We've also brought in respected people from Australia and South Africa to talk to the merino industry here. To see young people who have been through these experiences overseas start to come through in our industry has been fantastic for me.'

Ems and Will, meanwhile, have forged ties with a merino benchmarking group that meets three times a year to exchange ideas. 'They're all progressive properties with some really good thinkers,' says Ems. 'We always come back inspired. It's not just the facts and figures, it's being part of that group. Because farming can be a solitary sort of existence.'

So much, then, for the idea of the insular high country life. But what about more immediate neighbourly relations?

Glenmore's closest neighbour, with whom it shares the lake-end road,

is Godley Peaks. When the Scott family farmed the property, says Jim, 'it was recognised that you never left Tekapo without collecting their mail and papers for them, putting them in the mailbox and giving them a call. If they had a breakdown they didn't wait to ask, they helped themselves, and Bruce would call later and say, "I borrowed something from the workshop or the woolshed." That's the way neighbours around here have always operated. You can't just jump in the car and drive down to the dairy or whatever; you rely on your neighbours. When Godley Peaks was sold I rang the new manager after a bit of a snowstorm and asked, "Are you alright?" He rang back a few days later and said he'd never had a neighbouring farm ask that. I said, "Well, that's the way we operate here."'

Relations on Glenmore's other boundary, however, have historically been more strained. This is Defence Department land, and military manoeuvres have been a regular feature of Mackenzie high country life since the 1950s, when the army claimed rights on Crown pastoral leases.

Some of the army incidents at Glenmore were comical. In the 1970s, two armoured personnel carriers turned up at the homestead, lost and wanting directions.

'The guy looking out the turret called out "Mrs Mackenzie, I'm looking for the road to Glenmore", not realising he was parked in the Glenmore yard,' laughs Jim. 'We often used to have fun when the army did these exercises. Once they were dug in down by the larch plantation. I had an old tip truck and we decided to pay the platoon a visit, so we drove through the middle of their camp. The order must have been given, because every gun was trained on us as we drove in a big sweeping circle. We drove out, closed the gate, and left them wondering what had just happened.'

Shepherd Jack Holdom was once out at the Sardine Hut with Gerald when they heard whining noises overhead. 'Gerald stopped up the gully and yelled, "We're getting the hell out of here — they're shooting — and we're leaving the bloody Land Rover up on the ridge!" He was hoofing it.' By the time Jack caught up with Gerald he was a mile down the road.

When they got back to the homestead, Gerald got straight on the phone to the military camp at Balmoral. Jack heard him roaring, 'You can't go letting off ammunition without notifying the runholders!' The army yelled back that the live firing had been advertised in the paper, to which Gerald replied, 'Yes, but we don't get the bloody papers every day!'

In 1996 the Defence Department launched a bid under the district council plan change to take over 5000 hectares of Glenmore, including an area that had been declared a Site of Significant Inherent Value because of

RIGHT

Glenmore and Godley Peaks stations' shared letterbox.

the presence of some unique tarns. It was a tense and stressful time for the Murrays, and only started to come right following a face-to-face meeting with the top brass.

Dr Brian Molloy, a Lincoln University botanist with a long professional interest in the Mackenzie, made a speech to the Minister of Defence Paul East and the three heads of the armed forces that tipped the balance. 'He stood up and said, "Without hesitation you guys are the worst farmers in New Zealand. You can't just come in and take this land over and do what you want." That was the start of a turnaround, and the start of sensible relations,' says Jim.

Nevertheless, when the matter was decided in Glenmore's favour, the Murrays and Anne put 1000 hectares of the contested land into a QE II Open Space Covenant. 'It will protect it from military manoeuvres and from being blown to bits long-term,' says Jim.

Glenmore rams.

CHAPTER EIGHT

WOOL, BREEDING, SHEEP — AND GORDIE

Early March, and the Mackenzie stud tour has rolled into Glenmore, a cavalcade of Hiluxes disgorging farmers, potential buyers and New Zealand Merino marketing representatives onto the homestead lawn. It's the last stop on the opening day of a two-day circuit of the district's major merino properties, a long day that began at Black Forest Station down near Lake Benmore and is about to conclude with a catered dinner on the Glenmore greensward. But before the meal, everyone wants to inspect the Glenmore stock.

ABOVE
Tight parking in the Glenmore driveway.

RIGHT
Chris Bowman and Gordie McMaster, the two Australian sheep classers.

Will and his staff have penned some of the Glenmore stud's choicest rams and progeny at the base of the lawn. Drinks in hand, the guests wander from pen to pen, some wearing the same scrutinising mien you might see in an art gallery. Others, more tactile in their appreciation, run a hand through a fleece, rolling wool between their fingers or parting it to inspect the skin beneath. These are most likely sheep classers, high priests when it comes to breeding superior merino sheep.

Glenmore's wool is among the very best going, an achievement in which Will takes warm professional pride. In fact, in late May 2014, Glenmore won the Otago Merino Association's clip-of-the-year award, a South Island-wide competition. This fleece is the result of a team effort from sheep selectors, stock manager, shearers, wool-handlers and the wool-classer. 'We belong to a merino benchmarking group of eighteen properties and we're one of the top for wool production,' he says. 'Given that where we farm, our ewes are up in native country and spend half their winter standing in snow, that's not too bad. It gives you the drive to keep the stud going, because there's a lot of work involved and a lot of cost.'

It's a great distance from the scenario Jim faced in the early 1970s, when he and Anne founded the Glenmore stud.

WOOL, BREEDING, SHEEP — AND GORDIE

MIGHTY MERINO

Talk about going from hero to zero. In Spain during the fifteenth century, merino sheep were so prized that to export one was a crime punishable by death. In 1920s New Zealand, the sheep had almost been bred out of existence, victim of a shift in farming focus from wool to meat.

It was a mighty fall from grace for the first sheep breed brought in large numbers to New Zealand, the very basis of our national flock. Imported initially from Australia, and later from Germany, France, Britain and the United States, merino couldn't be surpassed for fine, soft wool. But the Spanish breed wasn't well adapted to New Zealand's wetter districts and heavier soils, proving particularly prone to foot-rot. It was probably inevitable that farmers would turn to alternatives. Today, the Romney is predominant.

In the South Island high country, however, the merino has always been king. Of the two million or so merino in New Zealand, this is where almost all of them live. Yet even in this drier climate, so much more like its ancestral homeland, the breed was never exactly 'optimised' for production. Small and scrawny, the early flocks only cut a single kilogram of wool. By contrast, the best flocks now cut somewhere between three and six kilos, a remarkable turnaround that you can chalk up to improved breeding.

Today the big controversy in the merino-breeding scene is about the role of science and statistics versus experience and instinct — having an eye for a good sheep. Estimated Breeding Values (EBVs), which describe the expected genetic performance of an animal — traits such as fleece weight, wool fibre diameter and body weight — have been latched onto by some as the holy grail. Others see them as a faddish irrelevance, computer geekery that misses the point. On a recent Mackenzie stud tour, one Tekapo farmer made the following pithy comment to underscore his antipathy to the growing use of EBVs: 'When the scorer becomes captain of your cricket team, then you're in deep shit!'

Jim Murray reckons the truth probably lies somewhere in-between. 'We're all striving for the ultimate sheep,' he adds. 'We'll never get there.'

Of the 140-odd bales of wool produced most years, twenty-five were sand; that is, were discounted due to too much fine grit having penetrated the fleece. (That wasn't necessarily unusual for the time; in its essentials, the story of Glenmore's wool reflects the evolution of New Zealand merino during the past few decades, as the box on page 154 elaborates.)

Jim explains that Glenmore's rams were all originally sourced from two properties in Central Otago, a different landscape and climate from the Mackenzie. 'The genetics just weren't up to the environment that we were trying to farm. I knew we could do better. So off I went to Australia, not knowing a hell of a lot about the Australian merino industry, but having seen one particular bloodline back in New Zealand that looked to be the type of sheep I was after.'

He returned with ewes and a ram from the Hazeldean stud in Cooma, New South Wales, a hardy line with a robust fleece. The plan was to breed rams tailored to the specific demands of the Mackenzie, with the immediate goal of packing wool on Glenmore's sheep, and a secondary aim of selling eighty to a hundred rams a year to similar properties.

'There was a lot of work in it. Ram selection, ewe selection, doing all the wool tests, weighing fleeces at shearing time, tagging at lambing for different bloodlines — a lot of work,' says Jim, who adds that sheep breeding proved one of the most frustrating things to get right. 'It's a moving target. In your mind you can have every aspect lined up, and for no apparent reason at all things just don't click and you end up with some very expensive dog tucker. Because it is certainly not cheap: these days, by the time you bring the semen in to do an artificial insemination programme you're looking at sixty dollars per ewe.'

He persevered, and enjoyed some success. But it took an encounter with a larger-than-life, often vexing, occasionally brilliant Australian sheep classer to kick Glenmore up to the level Jim wanted.

It's probably for the best that Gordon 'Gordie' McMaster is so short. The diminutive Australian sheep classer's in-your-face manner can rub people the wrong way. 'Gordie has had to be lifted off his feet and carried out of a few woolsheds in his time,' says Anne.

Jim met Gordie at the Christchurch Show in the early 1980s, where the Australian was guesting as a judge. He was colourful and controversial, but also clearly knowledgeable about merino sheep and how to get the best out of them. Compared with Australia, where merino is the dominant breed, the

ABOVE

The display board set up at Glenmore for the 2014 Merino Tour.

local scene was a step behind on issues such as sheep selection. And whereas New Zealand was too small to build a credible gene pool, Australian sheep classers had easy access to a large number of bloodlines for breeding. At the show Jim asked Gordie to visit Glenmore and a handful of other Mackenzie Country stations, an invitation he took up in 1984.

Gordie recalls he wasn't overly impressed by what he saw. By the early 1980s, the Hazeldean bloodline had established a few worrisome traits at Glenmore. 'They were short in the neck, very heavy, very blocky, wrong in the micron and bad "doers",' he says. As for the rest of the district, 'The sheep in the Mackenzie were lacking in guts — they were almost wrong for the country they were on.'

Like a pair of those rams, he and Jim locked horns. 'Jim would say, "Gordie, this thing isn't right, or this other thing isn't right." I'd reply, "Jim, that's crap, let's get onto the big picture. You want length and body staple, depth of body, wool cutting capacity — forget that other bullshit." Anyway, after the third year of my involvement, Jim was nit-picking and Anne turned to him and said, "Jim, either listen to him, or sack the bastard." She said it in front of everybody, and it was the greatest thing Anne could have done. After that things got a lot easier.'

The 2014 Merino Tour in the Glenmore garden.

ABOVE

The Glenmore garden during the 2014 Merino Tour.

Jim obviously has a different recollection. 'I kept saying to Gordie, "Look at this country!" It took him a couple of years to understand the environment we were farming. After that he was fantastic; he understood what we were trying to achieve. To have someone like Gordie come in from outside and objectively look at your sheep and say "You're doing this wrong" was brilliant as far as I was concerned. Initially, though, people looked at us sideways: "Why are you bringing in an external sheep classer?"'

One thing they agree on is that they were both ambitious beyond Glenmore's borders. Gordie: 'Jim told me, "I want to turn the Mackenzie Country into the best wool-growing area in New Zealand. Otago is now, but I want the Mackenzie to take that role." I said, "Righto, Jim, but first we need to change the sheep type."'

He persuaded the Murrays to visit Koonwarra Merino Stud, 300-odd kilometres south-west of Sydney just off the Hume Highway. Founded in 1947, the high-altitude stud had been at the leading edge of fine/medium-wool genetics for many years, winning awards across all the eastern states.

It was an eye-opening trip — and not just because of the sheep. 'We

began to appreciate the true value of these sheep classers, of which there were a large number in Australia,' says Anne. 'They could stand back and be objective, and they had experience of many types of stud in a vast range of country. In comparison, we had a small number of sheep breeders in this country, and our sheep were pretty inbred. Jim could see that to fix the faults and to aid the composition of the sheep, we were going to have to source bloodlines from outside New Zealand.

'Jim has always had a huge passion for wool, and he's been a perfectionist for getting the ultimate sheep,' she adds. 'He's been driven to breed something that will survive on a property with an eighty-inch rainfall at the top end, and just twenty-eight inches at the bottom, which is a difficult ask. You've got to have a wool that will stand that, so Glenmore is a tricky property in that respect.'

During the trip the Murrays 'fell in love' — Gordie's description — with a particular ram which had been the medium-wool champion at Sydney. They bought semen from that ram, and later returned to Glenmore with another ram, which they named Gordie in honour of their classer — 'he was a bit short in the leg and big in the backside, like me,' says Gordie.

Glenmore now averages 280 bales of wool per year, for only two bales of sand. The micron, a measurement used to express a wool fibre's diameter, is a couple of points finer. The station's lambing percentage has also improved — although, as Jim concedes, weather is a factor. 'Still, given a reasonable season, we've probably gone up fifteen to twenty per cent since meeting Gordie. Some of that will be due to additional feeding, but some is definitely down to the genetic breeding of the sheep. The Koonwarra sheep and wool type suit this country. They can stand the environment they have to live in here.'

Characteristically, Jim was keen to see the wider industry benefit from Gordie's expertise. He organised sheep-classing workshops at Glenmore for Mackenzie Country runholders, and encouraged Gordie to take his gospel to other parts of the South Island. Some might not have liked an Australian telling them that their merino operations weren't up to scratch, but Jim had no such illusions.

'Often the choice of sheep comes down to personal preference, which I think sadly has led some runholders into difficulties. Because it's not every type of sheep that will thrive in our environment.'

Eventually, Gordie would win them over — by 2000 he was classing sheep for forty New Zealand properties, fifteen of them studs. But in those early years he divided opinion. While personality played a part, Gordie's unconventional classing philosophy also set him at odds with many in the New Zealand industry. At a time when the scientific approach was in the ascendancy, Gordie favoured his gut. 'He was sceptical about the pure

ABOVE
Greta, Ben, Ems, Will, Angus, Jim and Anne at the 2014 Merino Tour.

OPPOSITE
Gordie McMaster: 'It's all in the face.'

scientific approach,' says Anne. 'For him it was about the visual appraisal. "It's all in the face," he'd say.'

'He's right, though,' adds Jim. 'You walk down the street in any town and you can size up any girl in seconds. It's the same with stock. You look for a nice soft nose, sign of a good and consistent wool type. Get a hard nose and you'll get a totally different wool type across that body. Then there are the fundamentals, like length versus height versus neck length. One of the first things Gordie identified was that our sheep had their heads coming straight out of the shoulders, no real neck — but you might get half a kilo more wool from a sheep with a neck that's fifteen to twenty centimetres longer. As for body length, Gordie liked to say, "This sheep is a chop short" — meaning it was less than ideal.'

The homespun aphorisms didn't charm everyone. When Jim introduced Gordie to Central Otago runholders, the Australian was asked to make the opening speech.

Recalls Gordie: 'Afterwards I was standing between Bill Gibson, the godfather of the New Zealand merino industry, and Martin Patterson, the owner of the stud hosting the event — both big fellas, and me in the middle, not much over five feet and built like a brick. And a professor from Lincoln University walked up and said, "Congratulations, Mr McMaster, you've just taken the New Zealand merino industry back ten years." I was about to punch the bastard in the nose when Bill picked me up on one side, Martin on the other, and carried me off — "Gordie, we think you need a drink."'

Gordie's getting a warmer welcome on today's stud tour. Before dinner comes out, someone hands Gordie a very large whisky. Earlier in the day at Balmoral Station, a Mackenzie stud breeder thanked him for being a catalyst for much-needed change. 'I thought to myself, That's as good a compliment as I can get.'

Will, too, has been complimented by

HOW TO CLASS A SHEEP*

You're looking for rich wool, with a surface that keeps the dust out. Check the 'nourishment', a coating at the very top of the wool that keeps it bright and clean, and that the crimp goes right to the end. If the skin is very elastic with a bright purplish tone, that's a sign of good blood flow. Look at the way the ram is standing. He should be 'correct', not 'hocky', square and straight and with a good deep body. Look at his face. A soft velvety face means soft wool. A long nose equates to a long body, a deep jaw to a deep body — these are pluses. Finally the horns. They shouldn't be tight to his head. He can get caught in a fence, and you need gaps to make shearing easy.

*With thanks to Chris Bowman, Australian sheep classer.

visitors on the quality of Glenmore's sheep. Like Jim, he has a passion for the stud side of the farm. 'We've got a number of clients and sell a hundred rams a year, so there's a responsibility to produce quality stock.'

'The reason why we cut so much wool, a lot of the credit has to go to Jim,' he adds. 'A lot of people chase micron — they go for finer and finer wool. Jim never did that. He just stuck to what he thought was the best type of sheep for this country, and he was right. We've been working off a very good base.'

But Will wants to cut his own track. He's farming in different circumstances to Jim, with new pressures and opportunities. Looking ahead, he can see that Glenmore will need a different kind of sheep. With prices for selling lambs to fattening operations in a slump, Glenmore may have to finish its own stock. More positively, Will sees real potential for merino as a dual meat and wool product. It's a big shift: traditionally merino hasn't been seen as good eating. But as a fine-grained, low-fat meat its time may have arrived.

The marketing may prove easier than the breeding, however. 'It's easy to put wool on sheep, but to put meat on sheep without losing wool is a tricky balancing act,' says Will. 'During the past five years we've endeavoured to free things up. The goal is more skin, more wool fibres, heavier fleece. The sheep are a lot plainer, but they're producing the same if not more wool.'

The big picture here is of diversification. Will and Ems are working a long-term strategy to reduce their reliance on wool sales, traditionally at least two-thirds of the farm's income. As will be seen later, there are significant changes afoot at Glenmore. A new kind of sheep is only the beginning.

ABOVE

Jim and Gordie McMaster, also known fondly as 'Laurel and Hardy'.

BELOW

The team in the kitchen preparing the Merino Tour dinner: Mary Stringer and Liz Scott.

Drafting ewes in the Joseph yards after the autumn muster.

CHAPTER NINE
THE GLENMORE WAY

'I can't hack this place, Gordie.' Johnny Bogue, aka 'Skippy', Gordie McMaster's Australian godson, had come to Glenmore aged eighteen for a bit of a 'straightening out' courtesy of Jim Murray. Gordie: 'I told his father, "He needs someone that will be tough on him. I'll send him to a mate of mine in New Zealand."'

LEFT

Dinner time at Waterfall Hut.

By the time Gordie visited Glenmore three months later, however, Johnny had had enough. 'We got in the car and drove down to the lake and he said, "I can't hack this place. I can't hack Jim. He's too bloody tough." I said, "Don't you let me down. I want you to stay at least twelve months. You get back there and you bloody work your head off."'

He did, and thrived. 'You couldn't have wished for a more willing or hard-working person,' says Jim. 'We've remained close ever since. Johnny now runs a construction business and he often talks about the standards he developed while he was at Glenmore.'

Johnny had had a taste of the so-called 'Glenmore Way', which sounds vaguely cultish but was simply the way that Jim Murray did things — exactingly, with minimal tolerance for slack standards.

Gordie: 'Jim's an absolute perfectionist. It's Jim's way or the highway.'

In assessing the potential of any prospective employee, Jim used to apply a quick test. After the niceties of introductions he would take a look inside their vehicle. If it was a dog's breakfast, Jim believed it spoke volumes of their approach to life, and therefore to work. Only once did this screening exercise fail him. Two months after he employed a farm worker whose car had been especially immaculate, and realising that the guy just wasn't up to scratch, Jim asked him: 'Where's that vehicle you brought to the interview?' The reply: 'That car? Oh, I borrowed it.'

A perfectionist, then, but far from perfect. Jim concedes to a host of stuff-ups during his farming career. Some proved costly. He once

THE GLENMORE WAY

put stock back too early into an area that he had poisoned with 1080 and lost 400 sheep. The poisoned carrot had thawed after being frozen for more than ten weeks.

'We used to have a pigsty behind where the present homestead is that I decided had to be removed. I figured the best way to do that was to burn it, which we duly did. What I hadn't allowed for was that there were cows and calves running nearby on the swamp area. We ended up losing half a dozen calves through lead poisoning — they were picking up the potash and consuming lead at the same time. You could definitely put that down as a stuff-up.'

Former neighbour Liz Scott of Godley Peaks recalls the day that Jim set fire to Glenmore's Top Block. 'The Glenmore men had all left to muster the Tin Hut Block when the Godley men looked across and saw the Top Block on fire, close to the hut and spreading. They all rushed to put it out.'

'I was packing up to move huts and emptied the fireplace,' explains Jim, 'and quite unconsciously I tipped the ashes outside — just sheer stupidity; a breeze came up and away it went. Our team came off the hill, and with the Godley men we extinguished the fire, using wet old jerseys to beat it out. If you were to think, This guy has never made a mistake in his life, then forget it, I've made plenty.'

Jim and Anne stress their indebtedness to the men and women who worked on Glenmore. 'The financial viability of the property totally depended on having loyal and hard-working staff,' remarks Anne.

The hardships of Gerald and Joyce's time required great loyalty from the staff. 'Jim

RIGHT

Jim Murray, at right, Johnny Wheeler, on trailer, and Will Murray in the Cass Valley during the autumn muster.

Calder and Cliff Moffat both made their mark at Glenmore during my father's time, and Bill O'Brien did much of the early building,' says Jim. 'Before them, Archie Mead, who managed the property for my grandmother, also left an imprint.'

He certainly did. Employed in 1914 to keep things ticking over until either Gerald or Bruce wanted to take on Glenmore, Mead got to work with a vengeance. In his first year he erected ten miles of fencing, went to war with the rabbits ('I had a fair kill,' he wrote in a journal), and built a suspension bridge over the Cass to get sheep to Godley Peaks for shearing. Somehow he also found time to manage the building of a new homestead for Mary, choosing a site near the head of Scott's Creek that was well sheltered and sunny, with a view of the water. 'In the spawning season, one could look out the window and watch the fish go up the creek.'

During the next decade Mead built up Glenmore's flock from 4400-odd to 7200 sheep, despite having only a single shepherd and a cow-boy for help and in the face of a particularly severe snowstorm in 1918. You can understand his chagrin, then, when in 1921 George Murray asked him to muster Godley Peaks, which by then had been taken over by Bruce Murray. Mead refused and quit. 'I went in there to manage Glenmore; I was not going to go back to mustering.'

Jack Holdom, who was Gerald's main shepherd and spent eighteen years at Glenmore, is particularly fondly remembered. Jack was Jim's early mentor and then his right-hand man following Gerald's death. When the 1967 snow hit, it was Jack who helped Jim pick his way through the aftermath.

Roger Mason was among Glenmore's longest-serving staff, with twelve years under his belt. David Grigg was another; now farming his own property in the Awatere Valley, Marlborough, he's returned for twenty-eight autumn musters.

Like David, Andrew Steven was employed during the early stages of Jim's intensive land cultivation. 'Jim insisted that the tractors were kept in immaculate condition,' says Andrew, of the 'Glenmore Way'. 'They were washed down every week — something that's unheard of on any other farm. He showed us how to keep our own vegetable gardens with the potatoes sown in a straight line. His fences are still straight and standing twenty years on. Now, as an employer myself, when I'm faced with a difficult situation I think, What would Jim Murray do?'

Pip Hunter-Weston, Jim's godson from Mt John Station, also worked at Glenmore. He remembers the colour-coordinated tools, hearth brush and shovel that Jim kept in the tractor cab — and admits that he does that himself these days.

'If you did anything wrong, Jim would let you know and make you do

ABOVE

Moving Helm's Hamlet to the Tin Hut, where it would become the woodshed.

BELOW

A step up from the old tractor — Glenmore's ex-army quad.

it again. He insisted that we take pride in ourselves. He expected every man to keep his quarters tidy, to cook and eat a decent breakfast every day, and to clean up before starting work. But he was always fair and you knew where you stood with him, and being on Glenmore was always a lot of fun. I remember Roger Mason, after the Glenmore hoggets had won the hogget competition, kissing each hogget as it was unloaded — this was after a few shouts at the pub from Jim.'

As well as Jim's demands, Glenmore workers had to be prepared to tough out routinely bitter weather conditions. Jim recalls one occasion in the 1970s when he and head shepherd Roger Mason, along with Bob Burnett, went up to check snow fences.

'It was a job that usually took three or four days. This time we were supposed to be upgrading a fence line above five thousand feet and we took the ex-army Chevrolet quad so we could bring all the fencing gear up in a trailer. I thought I'd assessed the weather, but I got it dead wrong, and on the second night we were hit by a twenty-centimetre snowfall, which forced us into the back of the old quad's military-style well-deck. Two of us slept either side of the well, the third curled up on the bottom. We camped there for ten days, upgrading the fence by day and with nothing but a sheet of canvas over us for shelter at night.'

Along with its permanent staff, Glenmore has been well served by a procession of contractors, inspectors, fencers, and part-timers, some of whom were wonderfully eccentric. Jack Skinner, an employee of the Pest Destruction Board, rode a white horse and had a wooden leg. 'You'd see him coming, this wooden leg flopping up and down on the side of his horse,' says Jim. Another pest board employee, Doug Helm, lived in a hut up the Cass called 'Helm's Hamlet' and had a technique of pouring meths through a thick hunk of bread smothered in Marmite to capture the raw alcohol.

Dozer driver Ian McPherson, by contrast, was all business. 'Ian drove a Caterpillar D6 and a 1952 Bedford truck called Blondie, and where Blondie didn't go was no one's business. He was a great guy for putting in tracks, just using his raw eye.'

Glenmore was also on the radar of young Lincoln graduates looking for work experience. Miles Carter was one. 'I don't think he ever did his hair. It was always blond and long and sticking up. We could never get Miles to wear boots. He always wore gumboots, even on the autumn muster — even to cook meals,' remembers Jim.

Occasionally, foreign workers stayed at Glenmore. Enrique was a cattleman from Argentina who spoke very little English and seemingly had a grudge against sheep. 'I can remember the boys had started to dip sheep through a spray race,' says Jim. 'Suddenly I saw two or three sheep thrown clean over the race. Enrique's patience had clearly run out: "Jeem, Jeem, these f—ing sheep won't go through here!" I explained that if he just slowed down then maybe they would.'

Jim prized loyalty above all else. One autumn, musterer Colin Neal developed an awful toothache at Tin Hut, about twenty kilometres from the homestead. He walked all the way back to the homestead, collected his car, drove to Timaru to see a dentist, then walked back from the homestead through the night to arrive at the hut at 5.30 a.m. in time for breakfast and a full day of mustering.

'Huge loyalty — and that's what we got from so many of the people who worked for us,' says Jim.

It's Glenmore's enduring mystery. For several years during the 1950s, the visiting shearing gang included a Maori shearer named Matt, who is remembered for two quirks. He ate his breakfast unnaturally quickly, as if afraid it might be whipped away before he could finish. And he carried an umbrella. Any time Matt walked to the shearing shed he would open his brolly, whether it was raining or not. A superstition? No one knew.

The shearing gang is a world of its own, even now. But in Gerald's day and the early part of Jim and Anne's tenure, the shearers who worked Glenmore were a law unto themselves. Today, you can find a professionalism among contract gangs — certainly in the gang that shears Glenmore. But back then, the runholder was at the mercy of his shearers' whims and after-work-hours appetites. Would they show? And if they showed, would they be of any use? Given their critical role at the money end of the farming year, shearers were in a position to call some shots.

'We had a particular gang of men for the first few years after our marriage, and they were a trial,' says Anne. 'I would have fifteen plates of eggs and bacon, toast and sausages sitting ready. On Monday mornings sometimes I'd find that only six or seven had arrived because the others were hung-over. In the end I was so livid that at lunchtime I produced the other plates with the cold bacon and eggs for the men's lunch. They got the message.'

It could have been worse: when Jim was a child, the shearers went on strike because they didn't like the food.

Shearers were hard-living characters, itinerant, indecorous, often indecent. In *Erewhon*, Samuel Butler describes the woolshed as cathedral-like, 'with aisles on either side full of pens for the sheep, a great nave, at the upper end of which the shearers work, and a further space for wool sorters and packers'. There was nothing of the cathedral about the language, however. At Glenmore, as on other properties, the shearers used a code word — 'ninety-nine' — whenever the runholder's wife or children visited, signalling it was time to keep the talk seemly.

Yet these men were also skilful exponents of their craft. Until 2005, Glenmore's sheep were blade-shorn, a practice that persisted in the Mackenzie Country and some other parts of Canterbury years after disappearing elsewhere.

'Because of the genetic improvements we've made to the sheep, coupled with the declining availability of blade-shearing gangs, we were forced into a rethink,' says Jim. 'But we've gone to cover-comb machine-shearing, which still leaves more wool on the sheep than a straight comb.'

In *Shear Hard Work* (2010), Hazel Riseborough notes that blade-shearing — using hand shears as opposed to powered tools — was always seen as better for animal welfare and survival in regions that experienced snowfall, because the blades leave behind slightly longer wool. Additionally, the atmosphere of a blade shed was noticeably calmer, and there was a flexibility that wasn't possible with machine-shearing, where the positioning of the downtubes determines the spacing between the shearers.

Hand-shearing appeals as a lost discipline, more skilful and more painfully acquired than the alternative. Yet when it comes to taking wool off merinos as opposed to other breeds, machine-shearing is an art in itself, writes Riseborough. 'It requires different gear, a higher level of concentration, more patience, an emphasis on quality over speed and a greater awareness of the product being harvested. A good merino shearer can shear crossbreds, but not all crossbred shearers, even the best of them, can shear merinos well.'

It's due to the combination of the wrinkled quality of merino skin, the fineness of the wool and its intensity — an area of skin the size of a postage stamp may contain as many as 15,000 wool fibres. 'Merino sheep are really more difficult to shear than coarse long-woolled sheep,' notes Samuel Bard in *A Guide for Young Shepherds* (1811). 'Shearing is a business which our farmers are too apt to hurry, but merinos deserve — and no doubt will meet with — more attention.'

Shearers didn't always have the whip hand. Before the advent of motorised travel, they had to hike to the high country, travelling in groups for safety when crossing rivers. Vance writes that they were sometimes forced to ford swollen waterways to be at an 'open shed' station on time. '"Open shed" stations picked their shearers from those present when the shed opened; there were no prior engagements and no jobs for latecomers. Today almost every shed is short-handed, but in those days more shearers than were required always turned up.'

Up until 2005, the shearing gang at Glenmore stayed on-site for a fortnight. In 1933, the same year he had the woolshed built, Gerald put up the existing shearers' quarters. It's used now for overflow guests and as a holiday let, and the shearing is done by a contract gang working out of Timaru. They arrive early in a convoy of vehicles, and head home at the end of the day, and they feed themselves.

Experienced blade-shearers could wield their razor-sharp tools with almost machine-like rapidity. 'Remarkable tallies were put up by fast blade-shearers. Gun shearers like Big Mick Radove shore fifty merino sheep before breakfast at Balmoral,' writes Vance, adding that inevitably there were accidents. 'The story is told that at Balmoral a rather stout shearer punctured his stomach with a shear blade. Pushing back the protruding entrails, he patched the wound and went on shearing.'

No chance of seeing that kind of accident at Glenmore today. On the evidence of the season of 2013, the modern contract shearing gang is all business. Even the shed hand's work is more skilled than you might imagine.

'There's a real technique to it, and you have to be able to make snap decisions,' says head hand Pania Warwick, a twenty-five-year veteran of the sheds. 'There are courses available these days. They go from level one to four — four-plus being where Jim is at with his wool-classing.'

It's also a year-round occupation — or close enough to it. Pania's gang is contracted out of Timaru to cover sheds throughout Canterbury. 'We go as far south as Omarama, and we go up as far as Arthur's Pass — that's our last merino shed, and there's a lake there so we can go boating and fishing in our downtime. After pre-lamb we have a couple of quiet weeks, and then we start

RIGHT

Greta keeping an eye on things as Ems works in the woolshed.

LEFT
—
'One person's opinion,' quips Jim.

into our main shear, which will take us through until March/April. We might have another quiet fortnight, then we kick off our North Canterbury pre-lamb —they lamb earlier than here. Then we're home for a week at the end of May, and then we kick into it again.'

It's physically taxing work. Aaron, a former fisherman from Nelson, is operating the wool press today. He's been a shed hand for seven years, never a shearer, but he's seen how hard it can be on the body. He points out one shearer working the blades whose back he knows is 'stuffed'.

During the busiest times, they're up before dawn and home around dark, for weeks on end. 'The thing about this industry is you have to be into it; there's no in-between,' says Pania. 'But if you are into it and you're a good shearer or shed hand, then you get to travel the world — New Zealand, Australia, the UK. It's a great lifestyle, lots of travel, you get to see a lot of country, meet plenty of people. The Murrays I know well because I've been coming to Glenmore for over ten years.'

The Glenmore shed tends to run like clockwork, she says. 'It's a big thing for us if the sheep are prepared well, and they're always well presented at Glenmore. It's pretty well known what they want done with their wool from the start.'

That specificity begins with written instructions from the buyer, New

Zealand Merino. In the Glenmore shed, these are kept close to hand, and include a long scene-setting preamble. 'Prices have collapsed since January 2013,' it begins bleakly. 'It is of major importance that growers hit their specifications. Premiums for superfines are almost non-existent and for mid-combing wools not much better. Fortunately, 50 per cent of New Zealand Merino clients have wool forward contracted, so it's important we work together. Missing contracts for VM [vegetable matter], length, soundness of style could be very costly for the grower. Make sure you know in advance of starting in a shed what contracts are signed and what is required.'

On the next page, Glenmore's contract requirements are exhaustively detailed, with tight targets set for length, micron range, VM, average yield, colour and style, along with an agreed quantity (191 bales), specified packaging and maximum weight.

As the wool-classer, Jim is responsible for everything in the shed from the moment that the wool comes off the sheep's back until the bales go out the door. 'If I have any concerns about wool quality or what's going on in the shed then I go straight to Pania.'

When the shed is at full hum, the classer becomes the quiet, forensic centre. He's an overseer, scrutinising the clip in its raw state to enforce those targets. For the first hour while he gets his bearings on this year's wool Jim is utterly absorbed at the classing table, checking against those specifications. Micron? The target range is 18.4–19.7, a pragmatic medium-fine wool. (See box, page 182.)

Once processed, the fleece is fed with others into a mighty Van-Gard High Country All-flex wool press and compacted into bale form, to be weighed and stamped ready for the truck. Glenmore's bales are destined for Merino New Zealand's Christchurch wool store, where they will be core-sampled against the contract specifications, then trucked onwards to Icebreaker or another buyer. For any wool not under contract, a small boxed sample is sent to Melbourne for auction.

As for the sheep, those 'productive units' will be dipped. Some farmers don't do a post-shearing dip. When Will tried that one year the sheep got lice — brought in, he assumes, from another farm by the different shearing gang he used that season. 'These sheep haven't got lice, so it's a preventative,' he remarks. They are then turned out onto the tussock blocks where they are left to lamb.

MICRON MADNESS

In 2013, at the Loro Piana Challenge Cup in Hong Kong, Mid Canterbury farmer Anna Emmerson produced the planet's finest merino wool, a 100-kilogram bale of 10.6-micron wool. It was a triumph, a total rout of the Australian competion, with Emmerson's 'ultra-fine' product destined to be turned into forty bespoke suits, each likely to cost more than $40,000.

There was a time when a 10.6-micron wool was almost inconceivable — in the first Loro Piana Challenge back in 1997 the winning wool was a comparatively hefty 13.4. But the worlds of fashion and outdoor apparel want finer and finer wool, and merino breeders and farmers have found ways to grow it for them. From an average micron of 20–21 in the 1980s, the New Zealand merino clip has dropped at least a couple of points. (By contrast, wool fibres used for carpets tend to be between 35 and 40 microns, and human hair around 60.)

To what end? The lower the micron, the silkier and softer the wool, with no issues of itching, and better breathability — for which buyers have traditionally been prepared to pay a premium. In the 1980s, the interest in chasing finer wools became arguably something close to an obsession, fuelled by some attractive premiums. During this period a single superfine merino ram sold at auction in Cromwell for $34,000.

Detractors say this 'micron madness' has come at the expense of breeding programmes. Inevitably there's a trade-off, they argue, and focusing too intently on producing finer and finer wools comes at the expense of yield, frame size and lambing percentages. Australian sheep classer Gordie McMaster reckons Otago in particular has suffered from chasing the micron prize — 'there's no guts in the wool'. He counsels a return to a more balanced approach. 'Let your country, management and climatic conditions be the judge, and above all allow Nature to run its course.'

For their part, the Murrays have steered a pragmatic mid-micron course, predicting that 'dainty' superfine sheep would never make it in the raw Glenmore environment.

On their way home.

LEFT
Ouf, surveying Top Block.

ABOVE
Barwoods trucks collect the Glenmore clip from the woolshed. George Murray helped establish this trucking company in the 1920s.

It's a pressured time, shearing, but there's always room for some levity, and a firm tradition of pranking between shearers and runholders. Jim once secreted some baby rabbits in the gang's van. As the shearers left for home all hell broke loose, shed hands shrieking and leaping from the van with rabbits flying in all directions.

Jim thought he'd got one over the gang a second time, on the penultimate day of the 2009 season. 'After they left for the evening I got all their moccasins and nailed them onto the roof. Then I pulled the tubes from the insides of the down-arms of the machines and disconnected the electrics. Next morning all of the shearers were affected in some way. When they headed off I was thinking, Great, this time I got the last laugh. The following day, the shearing shed was empty and I walked up to the classing table to class the remaining few fleeces, and it collapsed in front of me. The buggers had undone the bolts to within a fraction of it falling apart. So the last laugh was on me.'

THE GLENMORE WAY 185

Hell's Gates reflected in the waters of Lake Murray.

CHAPTER TEN

PESTS, PLAGUES AND POLITICS

One day in August 1997, Jim Murray received a phone call — he won't say from whom, but it was a local farmer — advising him to collect a 'package' from beside a culvert alongside the highway. That same cloak-and-dagger conversation, or a variation of it, had been repeated many times among runholders that month as, station by station, the rabbit haemorrhagic disease (RHD) was released into the high country. It was a highly illegal move — hence the secrecy — and one born of utter desperation.

In *High Endeavour*, Vance quotes Leviticus: 'And I will bring the land into desolation.' The reference here, of course, is to rabbits. From the day they appeared in the New Zealand high country, rabbits have come perilously close to delivering desolation, with the worst plagues taking their place alongside other dubious Mackenzie milestones such as killer snowstorms and floods. According to Jim, 1997 was as bad as it ever got.

'What the public of New Zealand don't realise is that in the year the calicivirus was released, the entire pastoral area, from Marlborough to northern Southland, was tipping on the edge of becoming a desert.'

Today it's almost inconceivable that rabbits were deliberately introduced to the Mackenzie Country, brought from Southland in the 1860s for sport. Even at the time, not everyone was blind to the risk. Vance highlights Charles Tripp, of Orari Gorge, whose first question to visitors was invariably, 'Seen any rabbits? Seen any gorse?'

Writes Vance: 'His son, Bernard, recalled a family holiday in Southland that took them past Lumsden. "We found the ground eaten bare. Even the railway line was undermined with rabbit burrows. My father was astounded at what he saw, and ran from one side of the carriage to the other, staring at the hordes of rabbits, who returned the stare."'

Tripp senior fired off an admonitory letter to the *Timaru Herald*. 'I fear that, unless most urgent steps are taken, our sheep runs will become useless . . . and will be taken up for rabbit warrens.'

By the 1880s it had become obvious that Tripp was right, and that the threat was urgent. A Rabbit Nuisance Act was passed empowering rabbit inspectors to compel landowners to destroy rabbits. To forestall an invasion from the Otago side of the Waitaki, the government built a wire-netting fence from the river opposite Kurow, up the Hakataramea Valley and over to the Tekapo River, then on to Lake Pukaki and up the eastern side of the Tasman Valley, terminating well beyond Mt Cook Station homestead.

All too little, far too late: by the end of the decade, rabbits were established in the Mackenzie, where they favoured the sunny facings in the warmest country. From that point on it was a running battle, marked by the establishment of rabbit boards and the use of ground and aerial poisoning, and farces such as the introduction of stoats and weasels.

Glenmore suffered as much as any station. From the 1940s, when the district was hit by its third major rabbit plague, much of Gerald's time was spent on rabbit control. Jim: 'They used anything they could, including gassing the burrows using phosphorus and poisoning using strychnine. We used to have two rabbiters' huts on the place, and I remember Gerald once saying that in just one winter a part-time rabbiter working out of the Cass Valley killed twenty-seven thousand rabbits.'

Rabbits were the worst but by no means the only threat to the land. Natural processes and harsh weather, overstocking, burning, deer, tahr and chamois, along with tussock-eating grubs, all contributed to the erosion of the Mackenzie Country. According to Vance, by the 1950s the sheep-carrying capacity of the district had been reduced by half. Among the public there was a growing impression that increased flooding of the plains was due to loss of vegetation in the high country.

Concerns about erosion in the district and elsewhere prompted soil conservation measures and stricter controls on burning. Planting, too. At the summit of Burkes Pass, the memorial erected to Michael John Burke was inscribed with a warning: 'O ye, who enter the portals of the Mackenzie to found homes, take the word of a child of the misty gorges and plant forest trees for your lives! So shall your mountain facings and river flats be preserved for your children's children and for ever more.' In the latter part of the twentieth century, Burke's successors finally got the sense of urgency. Conifers were planted in the loose scree fans of the high country, with major afforestation undertaken at Tekapo, Pukaki and Ashwick Flat.

But it was the rabbit plague that threatened most. By the end of the 1950s, an escalating poisoning programme appeared to have finally taken a chunk out of the problem. Jim credits the work of local pest control boards for this success. The Mackenzie Pest Destruction Board, founded in 1949 and responsible for some 360,000 acres of the south-western part of the district, recorded more than half a million rabbits caught and almost double that number killed by a phosphorus-based bait in its first two years of operations alone. Other boards proved equally efficient.

Jim was on the Tekapo Pest Destruction Board for twenty-seven years. 'These pest boards were set up in conjunction with the local farmers and the Ministry of Agriculture, so in all cases the local farming fraternity were board members alongside the bureaucrats. A rate on each property was set and collected, subsidised by the government of the day. In the Mackenzie Country, we had the Mackenzie board, the Tekapo board and the Pukaki board. They all worked particularly well, because there was that element of local interest.

'The pest board employed an inspector and two rabbiters as well, and land was set aside to build cottages for those people. We also bought our own plant, including a four-wheel-drive vehicle and a carrot cutter for the poisoning operations, and we put out tenders for the carrot supply.'

The standard pest board approach was to attack several farms in one concerted hit. 'You'd do a whole tract of land — in our case, say, from the Tekapo riverbed right across to Lake Pukaki — so you'd be getting a corridor of land poisoned in the same operation. You might have two or three properties involved, and they were required to make their land available.'

Like everything else on Glenmore, pest control was stepped up under Jim's regime. A rabbiter was employed, and anywhere between $40,000 and $80,000 was spent on poisoning operations per year.

Yet for all the carrots dropped and rabbits shot, every farmer in the Mackenzie knew that they were only holding the line, and that numbers could easily explode again. In that context, the decision taken in the late 1980s to disband local pest boards in favour of a single overarching entity was a terrible mistake, in Jim's view. 'The whole of the Mackenzie became one board, and then we came under the wing of a single South Canterbury board. And that was the demise of it — it became too big.'

Even as the rabbits returned, Glenmore was facing another potentially farm-killing scourge. Hieracium had been introduced into New Zealand soon after the First World War, secreted in horse-feed grains imported from Europe. Known commonly as hawkweed, it is a creeping, mat-forming perennial that grows in dense colonies which choke out all other vegetation. In the Mackenzie Country, where five species of hieracium took hold (*H. pilosella* being the worst), the weed invaded vast areas of pastoral lands that had been destocked for the summer months,

RIGHT
Drying rabbit skins.

PESTS, PLAGUES AND POLITICS

degrading pasture and stunting productivity.

Faced with these dual threats, Jim had to act. He reduced the flock by 2500, but that could only ever be a first step.

'We were all searching for answers, as we were losing a lot of grazing and needed to find a way to make that country grow more grass,' he remarks. 'We decided to put a direct drill through it — a triple disc, which was developed for just that type of country. That was when we engaged Richard Hutchinson, who, during one spring, sat on the tractor and direct-drilled more than a thousand acres with clover seed and applied the superphosphate through the drill. It was the start of a breakthrough to provide more grass in the most hieracium-ridden country. The following year we did another seven hundred acres. We ended up doing a bit over two thousand acres with a tractor and direct drill, putting in clovers and sowing it with superphosphate.'

Tackling the rabbits was even more daunting. By the 1990s, great swathes of the South Island pastoral country were struggling to cope with the invasion. At Glenmore, Jim was spending more money than ever. 'We were barely holding our heads above water. If we didn't do something we'd have been packing our bags and down the road.'

Enter the calicivirus. Farmers had already failed to sway the authorities on the wisdom of reintroducing myxomatosis, which had been released unsuccessfully in the 1950s. Instead, the government funded a $28 million programme to escalate poisoning operations and rabbit-proofing farms — to no avail.

'Farmers were going hell for leather financially and physically to try to control these things, but a couple of dry summers and the numbers just exploded again.'

Faced with a resurgence, RHD seemed to many farmers to be a no-brainer. Otherwise known as rabbit calicivirus disease (RCD), RHD had first appeared in the winter of 1983 in Jiangsu Province, in the People's Republic of China, then was rapidly spread by angora rabbits. Within nine months, fourteen million domesticated rabbits were dead, and the virus had reached South Korea, a first step in its steady transmission to forty or more countries, with most outbreaks occurring in winter or spring.

Among New Zealand high country farmers, it couldn't arrive quickly enough. They saw RHD as a last option — 'a virus to cure a plague', as a magazine once headlined. A group that included ten regional councils applied to the government for permission to import the virus from Australia, where it had 'jumped the fence' from a field-testing site.

The then deputy director-general of agriculture, Dr Peter O'Hara,

BELOW

Rabbit damage to a matagouri bush.

rejected the application. Despite public concern, it wasn't the risks of RHD to human health that decided the issue. Rather, the decision was based on three grounds: poor understanding of the disease, uncertainty about its effectiveness as a bio-control agent, and perceived inadequacies in the applicant group's arrangements to manage it.

Allowing these considerations to tip the balance against release was a 'significant departure' from previous decisions on importing bio-controls, noted a post-mortem on the RHD saga by the Parliamentary Commissioner for the Environment. 'Past decisions specifically acknowledged that [effectiveness] was highly unpredictable,' it pointed out.

It was entirely academic, in any case: within weeks of being rejected, RHD had been released — very likely it had been imported before the decision was made. Thus began the underground campaign. The Ministry of Agriculture and Forestry (MAF) made an attempt to put the genie back in the bottle, quarantining the property where the disease was first discovered, then investigating eradication. But it had spread too far, too fast, its effectiveness multiplied by the inoculation of captured wild rabbits, whose livers and spleens were harvested, homogenised and then applied to bait distributed over farms. RHD was released into the Mackenzie District, and Central Otago soon followed. 'There was a sort of ripple that went through the basin,' recalls Jim.

'It became clear,' wrote Peter O'Hara later, 'that the introduction and spread of the virus had been managed in a highly effective clandestine

PESTS, PLAGUES AND POLITICS

LEFT
—
Rabbit damage to young pines.

operation . . . MAF was forced to announce that no prosecutions for possession of the virus would be taken, in order to encourage farmers to provide the information needed to gauge the extent of the infected area.'

But Jim says there was no great master plan behind the Mackenzie-wide action. 'I think it was just a huge sense of frustration at the lack of understanding of what we were all up against and how devastating this was, and so any opportunity was going to be ruthlessly pursued.'

Upon uplifting his culvert 'package' — a rabbit infected with the calicivirus — he fed its organs through Anne's Kitchen Whizz, added water, and applied the mix to eighty tons of carrots already procured for a drop the next day. 'I had a quiet chat to the aircraft operator on that first morning and said to him, "Do a few runs over and above what's already earmarked for poisoning, would you?" — just to make sure that we got it spread everywhere.'

That first hit of RHD went through the Mackenzie like wildfire. Jim maintains that it is still very active and effective in the district, but there's a broad recognition that RHD cannot do the job alone. Recent Environment Canterbury (ECAN) blood tests of wild rabbits show three-quarters of them have developed some immunity to the calicivirus, and the warning has gone out to farmers that traditional culling methods will be needed to maintain control.

'Even with the calicivirus we were always sitting on a knife-edge, and no one knows how long its effectiveness will last. In two years' time it could be finished.'

Jim has no regrets about the initial release, however. Farmers, he stresses, were attempting to look after the land. 'I shudder, absolutely shudder, to think where New Zealand would be today if the calicivirus had not come in.'

The lack of understanding among the public and decision-makers about the RHD issue underscores something for Jim and Anne. Positions on the management of the Mackenzie have become polarised, says Anne, and few people appreciate that runholders have a stake in looking after the land. Farmers, meanwhile, have often been their own worst enemies.

'People saw vast pastoral grazing and forty thousand animals and they immediately assumed that farmers' practices had caused these changes,' she says. 'You'd lost your woody species, your tussock grasslands, and it was seen as being caused by farmers' poor management. It immediately caused a confrontational situation, "them and us", and I think that's been incredibly sad, and I think it's taken farmers a long time to get over a defensive reaction to any form of criticism. Nowadays the approach is more collaborative, with better understanding of the facts by the public.'

'But it *is* an emotive issue,' she continues. 'People love the tawny tussock, love the mountains, and they have a huge passion for the place. And because the land is leasehold, the public has a sense of ownership. You have to respect that, but also you have to realise that we have a right to the pasturage and are trying to make a living. We also have a huge obligation to look after that land; we're inspected and we have a stocking rate, and so there are lots of checks and balances. But it's all become hugely political.'

A pause. Outside the light is dying. Glenmore in the early evening softens, the tussock land glows honey-gold.

When she resumes, Anne says something quite unexpected. There was a moment late in their tenure at Glenmore, she reveals, when they seriously contemplated walking away.

The final straw was a dispute over Lake Alexandrina, which will be covered in a later chapter. 'We woke up one morning and the *Timaru Herald* had a headline about a moratorium on Glenmore farming. We'd fought the rabbits, we'd been in dispute with the army, and were looking down the barrel of a hieracium infestation, and we just thought, What are we doing here? How is it going to be for our children? And so we quietly went away and looked around the South Island for other merino properties, wondering what our options might be. And we just couldn't do it.'

Why not? 'Two reasons. There was no other area that had such good stock health. And one needed to have a passion, and Glenmore was it for us. This was the place that we felt passionate about.'

Cousins Harry and Angus, always on the lookout for tahr.

CHAPTER ELEVEN
THE CHILDREN'S EXPERIENCE

'Four hundred metres, right through the neck.'

Angus Murray, Will and Ems's eldest, has just shot a bull tahr, and you can hear the pride in his father's voice as he relates the vital statistics. A seven-year-old boy, a .22-250 rifle, and a single shot at long range for the kill.

LEFT

The next generation: Greta, Ben and Angus.

The typical Mackenzie high country station has seen change upon change during the past two decades, but in some essentials it remains unchanged. For a farm kid like Angus, life is still that thing that mostly happens outdoors, up a river and among the mountains, or riding on the back of a ute with the dogs. As Will puts it: 'For every boy growing up on a farm, his farm is the most special place in the world.'

His own childhood was defined by its freedoms and duties. From an early age, Will and his sisters Kate and Pip enjoyed a right to roam that would amaze contemporary urban children, exploring the more accessible parts of the farm on foot, and further afield on cross-country skis or by bike. 'We could pretty much do what we wanted, go where we wanted.'

He hunted rabbits with a slug gun and, with the girls, made spears out of manuka hill-sticks and old bits of wire and went after the fat trout that came up the creek from Lake Alexandrina to spawn.

That last activity got them into trouble once, he recalls. 'One day we had a heap of friends here and we'd hauled out quite a few fish — possibly eight or ten, and they were big fish, the kind prized by fishermen. We had them all lined up on the bank when a ranger turned up. He knew who we were and he just started ripping shreds off us. Pip began crying because she was sure we were off to prison.'

Thinking they'd escaped with a stern talking-to, they all headed home, unaware that the ranger had beaten them there and was relaying everything to Jim and Anne. 'We got home, soaking wet and covered in fish, and burst in the back door yelling to Mum about this bloody old ranger, and there he is, standing in our kitchen!'

They were a tight trio, with only two-and-a-half years between Kate the eldest and Pip the youngest. Anne contrasts the gulf between Jim and his much older siblings. 'Our three were a formidable unit and very protective of each other, and to this day they are very close.'

'A good friend of all of ours still calls us the "Murray Mafia",' says Pip, who is married to Hamish Smith, Will's great mate, who once worked as a shepherd on neighbouring Godley Peaks, and who now farms just down the line in Omarama. 'Living rurally we spent a lot of time together, and that continued in the holidays in the years after we went off to boarding school. We were free-range children, who used to tootle off together and build tree huts or head over to the sheep yards to see Dad.'

They were classmates, too, studying their Correspondence School lessons in a dedicated classroom upstairs under the tutelage of the shepherd's wife. 'We were naughty kids,' says Pip. 'One particular teacher we really didn't like. She went home for lunch once, and when she returned we were standing on the roof, pelting her with pine cones.'

All of the high country families with school-age children got involved in the Correspondence School camps and sports days. The home-based syllabus also allowed the Murrays flexibility to attend field days and to take family holidays based around the farming schedule.

Discipline tended to be Anne's domain, although when Jim raised an eyebrow everyone took note. 'He just pretty much had to look at us,' says Will.

Will, an accomplished practical joker, pushed his luck at times. 'I locked Dad in the petrol shed once, which I thought was funny. He wasn't that upset, but when I let him out he grabbed me and threw me off the bridge into the little creek. It was only a foot or so deep, but I thought, Bugger this, I'll have you on, and acted dead. Well, that backfired because Dad got a hell of a fright.'

As they grew older, they began to understand just how hard their parents worked. Jim was often absent until late at night with farming duties — 'I remember going for days without seeing Dad,' says Kate. Anne was constantly cooking for great hordes of workers, or helping Jim around the farm.

'Mum is of that generation where she has devoted her life to supporting Dad and Glenmore,' says Pip. 'Her focus was on the farm and the children and that support role. She was stoic; she just did it and she never complained. That was common practice in those days.'

The children, too, were all expected to help out on the farm, particularly Will. At what age did he know that he wanted to farm Glenmore? 'When I was about one,' he answers. 'I often wonder what I would have done if I hadn't been from a farming family, but that wasn't a decision I had to make. I just knew I wanted to farm.'

By the age of eight or nine, Will was helping Jim during shearing or at crutching time. The latter put him in something of a bind. 'For some reason Dad liked to crutch on the same weekend that the duck-hunting season opened. I loved my duck hunting, so I'd be looking out the window the whole day. After we finished, there'd usually be half an hour or so to grab your gun and have a shoot before dark.'

When Jim put in the border dyke irrigation, Will was roped into making the concrete, throwing shingle into the mixer until he was 'absolutely buggered'. He tagged along on barn-building duties, too, although here he was less useful.

'Being a boy I picked up every tool, broke drill bits, that sort of thing, and got growled at. Once I got a pocket-knife for a present and I tested it by cutting every string on every hay bale in the barn, then for good measure slashed the temporary silo liners to find out what was inside. Jim later found tonnes of oats lying on the ground. That one was a bit more serious. But it was all good fun, and Glenmore was a fantastic place to grow up as a young kid.'

He was only six when he experienced his first autumn muster. Too small to walk the hills, he spent three nights camping out with his father and the men at Memorial Hut and the Tin Hut. 'I mostly helped the cook,' he says. 'I remember one day we were mustering the Tin Hut Creek and the cook told me, "I'm just going to check this side creek out; I'll only be gone an hour." He left me sitting on a rock, but he must have found some sheep and got held up. When he didn't come back I got really nervous. About two hours later Jim came back on his beat and saw his six-year-old down in the riverbed, bawling his eyes out.'

At the age of nine, Will's life changed dramatically when he was sent off to boarding school — as Kate had been before him, and Pip after. He remembers the sense of dread he felt on the eve of heading off to Christchurch's Medbury School.

'I hated the idea of going so much I decided to run away. I planned it all in my mind. I was going to pack a bag of spaghetti and baked beans and walk up the valley to Waterfall Hut. I was all set to go that night, but then I fell asleep and the next thing I knew it was morning and I was being thrown in the car and taken to Christchurch.

'We were all reasonably shy kids who had grown up on Correspondence,

RIGHT

Greta, Ems, Ben and Angus.

and the thought of leaving Glenmore and the security of home was just sickening. Once we got there it was fine, I enjoyed school, my new friends and all the sport, but every holidays I'd be counting down to when I had to go back and it was heart-wrenching, especially in the early years. We weren't allowed to see our parents for the first month, and then we probably only saw them once a month.'

The start of the holidays, by contrast, was a relief, like breathing again. Pip: 'I absolutely loved it. You'd wake up that first morning back, hear the birds singing outside, open your window and think, Phew, home.'

Hard-working as they were, Jim and Anne always made time for memorable family trips. 'Jim and I wanted the children to have a broader range of experience and to have more fun family time than Jim had ever had,' says Anne.

These holidays were a continuation of the theme of their Glenmore lives, which was all about adventure and enjoyment of the outdoors. They spent weekends with farming friends camped in musterers' huts on different properties, and there were regular sailing, skiing and skating parties on farm tarns. The children were hurled around, sitting in beer crates attached to long pieces of rope, or behind a farm four-wheeler, wheels spinning on the ice.

At other times they went waterskiing on the Mackenzie or Central Otago lakes, or jetboating up West Coast rivers. Kate has vivid memories of Jim taking the three children over the Copland Pass alpine crossing, from Aoraki/Mt Cook National Park to the West Coast. The Murrays maintained a bach

LEFT

The children off to Pet Day at Tekapo School.

RIGHT

Will, Pip and Kate, in their teenage years, relaxing on the front verandah of the homestead.

BELOW

Angus at the end of a long day.

at Wanaka for skiing holidays, and another in the Marlborough Sounds for summer sailing. Closer to home they skied Roundhill, the Tekapo field they'd helped pioneer. And when Pip turned ten, Jim and Anne took the kids for two months to Europe. It was an eye-opening trip, and lit a fire in all three to travel as young adults.

Glenmore may have been geographically isolated, but their childhoods weren't insular, stresses Pip. 'We were lucky that Mum and Dad exposed us to so many different things.'

'All the experiences and responsibilities they had at a young age led them to have a quiet confidence in themselves and a lifetime's sense of adventure,' adds Anne, who emphasises a common sense of resilience and independence among her three children. 'Compared to Jim's childhood, they were very lucky children. They all went to university and had their own vehicles and travelled widely. In saying that, they also all knew how to work hard, to take responsibility and to look after and appreciate what they had.'

As for her grandchildren, she adds, they are living a different kind of Glenmore life again. 'Will and Ems spend more time doing all these

ABOVE

Father and son at the end of mustering: Will and Angus make for home.

recreational activities with their children than we did. Angus, Greta and Ben have definitely experienced the skiing, fishing and tahr shooting at a younger age and more often than ours did. They all go to school in Tekapo each day, and have attended play centres, too, so they've had more social contact.'

The Murray children have a sense of independence that some town kids wouldn't recognise. As soon as he can wolf down breakfast, Angus is out the door, onto his mini motorbike and up a farm track.

Ems recalls a morning when she was helping in the woolshed and heard Angus blat past, off on some adventure. An hour later, concerned that she hadn't heard him return, she went looking and found him driving home on the main road, dragging a tree behind the bike, followed by a couple of concerned tourists in a rental car.

'His story was that he'd taken his axe up the property, chopped down a tree he wanted, but then got the rope wound around the back tyre. He couldn't fix it himself, so he waved down these tourists — he told me later they "talked funny" — and they were following to make sure he got home safe. I didn't know whether to tell him off for driving on the road, or congratulate him for finding a solution to a problem. In the end, I just decided to be proud of his resourcefulness.'

ABOVE
—
Anne and Jim's younger daughter Pip with her husband, Hamish Smith, and children, Harry and Suzie.

'There's a very real need for resilience growing up in this environment,' she continues. 'They sometimes need two to three layers of merino clothing, woolly hats and snow boots just to go out to feed the chooks! But then there's staying in huts, sleeping under stars and cooking on open fires, helping to shift sheep, driving the jetboat and the tractor, searching birds' nests for eggs, shooting rabbits and ferrets, feeding the animals — they all thrive on the responsibility of having animals to take care of. It's a pretty special childhood for them, a special way of life.'

Anne watches how the high country life is slowly shaping her grandchildren. 'As with all of us, living at Glenmore has a huge impact on their lives, in terms of their identity, self-sufficiency and sense of belonging,' she says. 'I can see it through all the generations alive today. Aged eighty-four, eighty-one and seventy-nine, Gendy, Erf and Pat still think of Glenmore as home. Kate and Pip have always had a great desire that their children should have a chance to experience life at Glenmore, too, so the cousins all spend time there. Glenmore has had a huge imprint on all of us.'

Is Glenmore still home for Pip and Kate? 'It becomes part of your blood,' says Pip. 'You have such strong memories of the smells and the sounds of the place. And the high country itself is a powerful place; if you're lucky enough

THE CHILDREN'S EXPERIENCE

to be brought up there, it becomes part of you, I think. Glenmore still felt like home to me right up until I got married and had my own children. Omarama is home now, but coming back to visit Glenmore is lovely and familiar and beautiful.'

For Kate the connection is more powerful. 'For me there is still a huge sense of Glenmore being home, a real sense of belonging to it.'

She regularly brings eight-year-old Milly up from Wanaka for visits, showing her the Glenmore she knows, opening her eyes to the life of a working farm. 'It might be something as simple as going out to the paddock to watch Will baling up hay — that's something she's never seen before, and I get huge pleasure out of sharing things like that with her.

'In the end I just feel incredibly fortunate to have grown up at Glenmore, and for all the experiences we had there. It's something I hold very dear, and I'd like it to continue with this next generation, to see them have that same sense of adventure and a love of the outdoors.'

ABOVE

Kate's partner Paul Nicol, with Milly and Kate.

RIGHT

Three generations: Angus, Will and Jim.

Mish Mackenzie, Gus Mackenzie and Angus Murray. Ailsa Stream is in the background.

CHAPTER TWELVE
HUNTING TAHR AND SAVING STILTS

Glenmore Station begins at a lakeside and ends on a mountain top — feet in the water, head in the clouds. Two animals define the fringes. In the littoral zone, the rare black stilt, a national rehabilitation project. Up in the craggy tops, the elusive bull tahr, a trophy that people will cross the world to hunt. In a sense these animals also define the distance between old and new Glenmore — one an object of official protection, ring-fenced and monitored; the other the focus of long-standing high country tradition.

ABOVE
A bull tahr with a nanny.

Nothing's ever that clear-cut, however. These days, tahr hunting on Glenmore land is handled for the Murrays by two sanctioned commercial guides. Managing public access became too burdensome, says Anne. 'We had up to 350 parties going in every year. We gave the hunting rights initially to Walter Speck, a Swiss guy from Tekapo, who knows where the animals are and can manage it. We simply clip the ticket for the use of the huts and so much per animal shot.'

Yet hunting tahr remains a part of station life for the Murrays. Will grew up shooting in the hills. He was fifteen the first time he went hunting with Jim. 'I couldn't sleep, I was that excited. From then on I'd go out in the weekends no matter what the weather was like. I used to live it, breathe it. I did a lot of hunting on my own, which I loved.

'For me the buzz was getting out in the hills and having the freedom to go anywhere I wanted. Mustering, you had to go a certain route; but when I was hunting I could explore all the nooks and crannies. That's how my climbing started, I just kept going higher and higher and one day I got to the top of Hell's Gates on my own. I was seventeen and no one believed me. After that I gave up my rifle and picked up climbing.'

At seven, Angus is an even more enthusiastic hunter. 'The little ratbag shot his first bull tahr when he was six,' says Will. 'There was a fair crew of us walking through this area and this bull bolted past, and Angus shot it, right through the head from a hundred metres. Another time we were driving up the Cass, me and James, who's married to Ems's sister, with his son Arthur and Angus in the back. I said to James, "You quite often see tahr bulls walking across the riverbed", and Angus in the back is screaming, "There's one!" Sure enough, there was a big bull. By the time we'd got set up this tahr must have been four hundred metres away, but Angus slotted it. He's got the bug now.'

The first tahr were introduced into New Zealand in the first decade of the twentieth century, thirteen of them liberated in the Mt Cook area, followed by a smaller number of chamois. Transplanted from Nepal for recreational hunting, they did so well in the tussock grasslands of the Mackenzie that numbers exploded. Today the tahr population is subject to a Department of Conservation (DOC) control plan, a recognition of the damage they can do to the vegetation and soil of the high country. Bull tahr, however, are all but venerated by hunters.

Walter Speck, one of two hunting guides working on Glenmore, says the magic lies partly in the tahr's Himalayan origins. 'Tahr is an extreme mountain animal, regarded as the best climber in the animal kingdom, better than chamois, and able to climb closer to the rocks than ibex because it has smaller horns. That has a certain appeal to mountain hunters. Also, a mature bull tahr is a stunning-looking animal, with a blond mane and a dark rump. There's an appeal, too, in the scarcity of the animal — you can really only hunt tahr in Nepal or in New Zealand.'

His clients tend to be from Europe and fall into two camps. Some are trophy collectors — they want one of everything mounted on the wall. The majority, however, are more like Will, keen on the experience of hunting in the mountains. 'Glenmore is very rugged and mountainous, and it can be a challenge for some of these hunters. They're not like musterers, and this kind of terrain is a novelty for them. People have been living in the European Alps for two thousand years; there are game trails, little tracks to follow. Here, once you leave the valley floor it's all untracked, and it's wild and very steep country — which is part of the attraction for these visitors.'

Spotting tahr in such a landscape is a challenge. Summer, when the bulls head up to the tops, is particularly tricky. Around April, they mob-up in preparation for the rut and become more visible. In spring, tahr descend to the valley floor to feed on fresh grass.

Walter has been guiding on Glenmore since 1996. 'Your eyes become trained — like a proofreader spotting spelling mistakes on a page. You also have traditional information about where the animals are likely to be found. You try to find them in the morning, when they are feeding. During the day they move higher and higher up into the mountains. When they get up into the rocks they bed down for the day, so between late morning and mid-afternoon they become very difficult to see. But you may spot that early movement. If it's a clear day, they glisten under fresh morning sun.

'They are very careful animals. They will look at the land below them before they come down. Tahr will stand motionless for ten minutes, staring down. If they see something suspicious — could be a glistening on a pair of binoculars or a rifle barrel — they will keep watching while they climb back up. They don't bolt like deer; they keep eye contact. For them, we're like wolves, and they will try to pull a hunter up into the rocks to disorientate him. If they spot something suspicious in the morning they will feed in a different spot that evening.'

Tricky as tahr are to hunt, Walter assures his clients that they have a realistic chance of getting a trophy animal and that they should at least see some bull tahr while at Glenmore — peace of mind for which hunters are prepared to pay a daily fee for his guiding, plus a handsome premium if they bag a trophy. On DOC land, by contrast, hunting pressure is far greater, especially as a result of heli-hunting. (Glenmore has suffered from airborne poaching. Will once took a photo of a chopper with a tahr hanging from it, then posted it to the aggrieving party, along with a letter from his lawyer; a cheque came back.)

RIGHT
Young bull tahr in the Cass Valley.

There are signs that the number of bull tahr is declining. As well, recreational hunters shooting younger bulls have created imbalances in the population, according to Walter, who only stalks mature animals. 'If you do it on a commercial basis you are far more selective about what you shoot. You manage it, or end up with hardly any trophies,' he says. 'Basically, we are farming free-ranging animals.'

— ✕ —

That's one take on animal conservation; Dr Ray Pierce comes at it from a very different place. Ray, who these days is based in Queensland, from where he leads a project to restore the ecosystem of the Phoenix Islands in Kiribati, is the original watchdog of the black stilts at Glenmore. He arrived on the Murray's doorstep in the winter of 1977, an Otago PhD student looking for a base from which to study black and pied stilts for his thesis. It wasn't unfamiliar turf.

'I was introduced to Glenmore as a schoolboy, when I would go camping with my parents at Lake Alexandrina,' he says. Later, as I got interested in birds, I got to know the area quite well, because, apart from black stilts, there were crested grebes on the lake, black-fronted terns and plenty of others. But it wasn't until 1977 that I lived there. I was doing a comparative study of pied and black stilts; the pied stilts were thriving in New Zealand and the black stilts were bombing out, and I wanted to look at aspects of their feeding and their breeding biology, and the impact of the introduced predators that are so conspicuous around Glenmore — the feral cats, ferrets and so on.'

Jim and Anne set Ray up in a hut on the property, where he camped on and off for the next decade, living on the smell of an oily rag. 'The Murrays were very welcoming. We had our moments, but I think they were pretty chuffed to have a student doing studies there, and they were always happy to have an extra pair of eyes around. The pest board gave me ammunition, thinking I was controlling rabbits, but really I was just cropping them to bait my traps and for food for me and the dog. Jim and Anne had put the electricity on to the hut, there was water, a wetback — it was a home away from home for me.'

Says Jim: 'I remember one of our first meetings with Ray Pierce. He came into the house with a long face. He said, "I've buggered one of your gates. I was riding my motorbike up the Cass Valley and I was so busy watching the birds that I ran into the gate."'

You can understand why Ray was so preoccupied. The endemic black

ABOVE

Black stilt expert Dr Ray Pierce.

stilt (*Himantopus novaezelandiae*) is the world's rarest wading bird, with an estimated wild population of eighty-seven adults. It's suffered a precipitous slide towards extinction: at the time of first European contact, black stilts were commonplace, especially on the braided riverbeds of Canterbury and Otago. By the middle of the twentieth century, however, it was rare to see a black stilt anywhere other than in the upper Waitaki Basin.

Ray estimates that the regional population was somewhere between 500 and 1000 in the 1940s, but within two decades had collapsed to fewer than 100 birds. Meanwhile, the *arriviste* Australian pied stilt, which had been virtually unsighted in early colonial New Zealand, went from strength to strength. By the time Ray was into his work at Glenmore, the two species had traded places — twenty-three black stilts recorded in 1981, versus an estimated 30,000 pied.

Glenmore was among the last bastions, with black stilts still to be found nesting in wetlands beside Lake Tekapo and up the Cass River. Observing both stilt species, Ray arrived at an understanding of why the black stilt had fared so badly while the other had prospered.

'Introduced predators impacted far more severely on the black stilt than on the pied stilt, which as a recent immigrant from Australia had evolved in the presence of ground predators. I could see lots of differences in their ecology and behaviour that made the black stilts more vulnerable, like the fact that they nested on riverbanks whereas the pied stilt would nest on islands.' Black stilts, he added, 'were slower breeders with a longer incubation and fledging period, and had distraction behaviour aimed at aerial predators, while the pied stilt knew how to lure ferrets and other nasties away from their nests.'

Loss of habitat had exacerbated the black stilt's plight, especially in lowland rivers where gorse, broom, willows and lupin choked up the banks where the birds prefer to breed. A further risk is now posed by cross-mating between black and pied stilts.

But extinction has been averted. Since 1981, DOC has mounted an intensive rearguard action, based around a captive breeding programme near Twizel in the Mackenzie Country. Reared in captivity for their most vulnerable early months, birds are released at various sites around the basin.

At Glenmore, conservation efforts are centred on a large deer-fenced exclosure on DOC land near the lake. Towards the end of his PhD, Ray became technical adviser to the project's funders, Forest and Bird.

'Jim and Anne provided a lot of logistical support with their vehicles. I remember when we were building the exclosure and they were away at Wanaka, we got the tractor stuck in the creek. One of their hired hands and I borrowed Glenmore's bulldozer, but then that slid off its tracks into the creek. We spent all day getting it out, and then retrieved the tractor, and finally got everything back in the shed. But later Jim arrived home and noticed all these strange vehicle tracks coming and going. He figured it out straight away: "Bloody Pierce, he stole the tractor and the dozer!"'

During Ray's time at Glenmore, he got to know a few individual birds well, including one he named Lucky. 'When Lucky was a week-old chick, there was a tremendous snowstorm and his mother died. The father was still there but I decided to uplift the chick because he was so weak. I took him back to the hut inside my jacket and kept him in front of the fire for a week. I'd go up to the Cass River bridge with a drift sampler and collect thousands of mayfly larvae to feed him in front of the fire. When the weather warmed up a little I noticed that his father was still down in the exclosure, and so I took the chick down and let it go. It wasn't clear what was going to happen, but the father eventually saw him and he was like, "Good God, where did you come from?" A cloud came over and the temperature suddenly dropped and the two of them just ran at each other, and the father brooded that chick. We saw Lucky for years and years after that.'

There have been heartbreaking moments, too. One weekend when Ray and the Murrays were away someone broke into the exclosure and stole eggs — worth $80,000 on the black market of the day, it was said.

Today, Ray is only cautiously optimistic about the black stilt's future. 'As a species it's very dependent on management. But it's not just the black stilts we're losing. In the Glenmore area, which has been a stronghold for several species, we've seen a decline in black-fronted terns, black-billed gulls and banded dotterels. They're all being hammered by introduced predators.'

RIGHT
Black stilts.

The new homestead, built in 1990 after the fire.

CHAPTER THIRTEEN
THE FIRE

Each generation of Murrays at Glenmore has had its bitter pill: the snowstorm that kills half of a young man's flock, the plague of rabbits that threatens to eat the land from under him. The high country does everything on a grand scale, including adversity. For Jim and Anne Murray, however, their lowest moment was man-made.

It was 27 November 1989, and Kate and Pip were both home studying for exams. After breakfast, Anne drove Jim and the girls up the Cass to retrieve some stray cattle, then set off back to the homestead.

Pip: 'I remember we were bringing the cattle through the Cass Gorge when we smelled smoke. Little bits of ash started to fly past.'

Anne returned to find the homestead on fire. She flew back to collect Jim, leaving the girls to bring the cattle home. By now the house was fully ablaze, smoke billowing into the big Mackenzie sky. Built of greywacke stone, the walls produced an effect like an oven, encasing the heat until it reached such a temperature that each room in turn exploded and the roof collapsed.

Flying overhead for the Helicopter Line, Gavin Craig saw the scene and on his own initiative returned to base for his monsoon bucket, then began flying sorties between the blaze and Lake Alexandrina. Jim, meanwhile, was frantically trying to get a high-volume pump to the swimming pool. When the fire brigade arrived and struggled to get their pump working from the creek, he yelled to 'bring the damn engine across the lawn to the pool'.

None of it made a difference. Anne remembers their contrasting first reactions. 'When I drove down and saw the house was on fire I became completely calm, which is the way I tend to react in an emergency. Nobody was hurt — the family was safe — so I looked to the practicalities. When Jim saw the house ablaze he had an immediate reaction. He straight away comprehended the devastation and all the implications — a hundred years of family photographs and documents, his grandfather's farming journals, all the merino stud records, our possessions, all gone.'

For the first time in their lives, the girls witnessed their father lost to his emotions. 'Seeing Dad with tears rolling down his face . . . it was full-on, a very hard time,' says Pip.

In the aftermath, the chief fire officer from Timaru blamed a faulty three-pin plug for the fire. Jim had another theory. Earlier in the week, builders had been working on an alteration to the kitchen. He speculated that a nail had been driven through a wire in the ceiling, causing it to arc. While eating breakfast that morning he'd been frustrated by constant static on the radio — perhaps evidence of a short-circuit in the wire.

Whatever the cause, the homestead was lost. At boarding school, Will got the news in a halting telephone conversation with Jim. 'He was talking very slowly and coughing: "Will, I have got some bad news." Big silence. "No one has been hurt." He's coughing away — "Dad, why won't you tell me what's going on!"

'I came back for a few days. It was horrible, and the worst thing was the

RIGHT

One of George Murray's farm diaries, rescued from the charred remains of the house.

The SHEEPMAN'S DIARY.
1903.

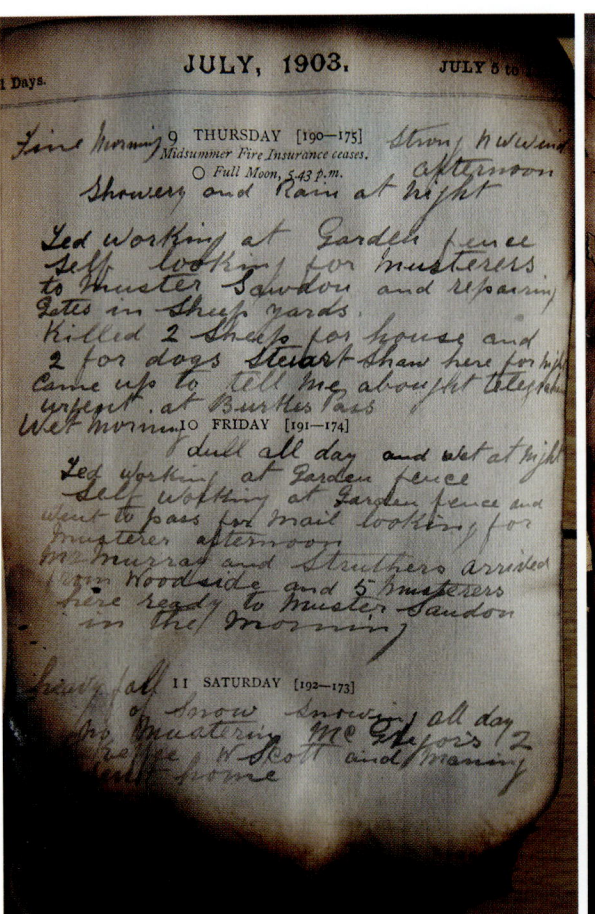
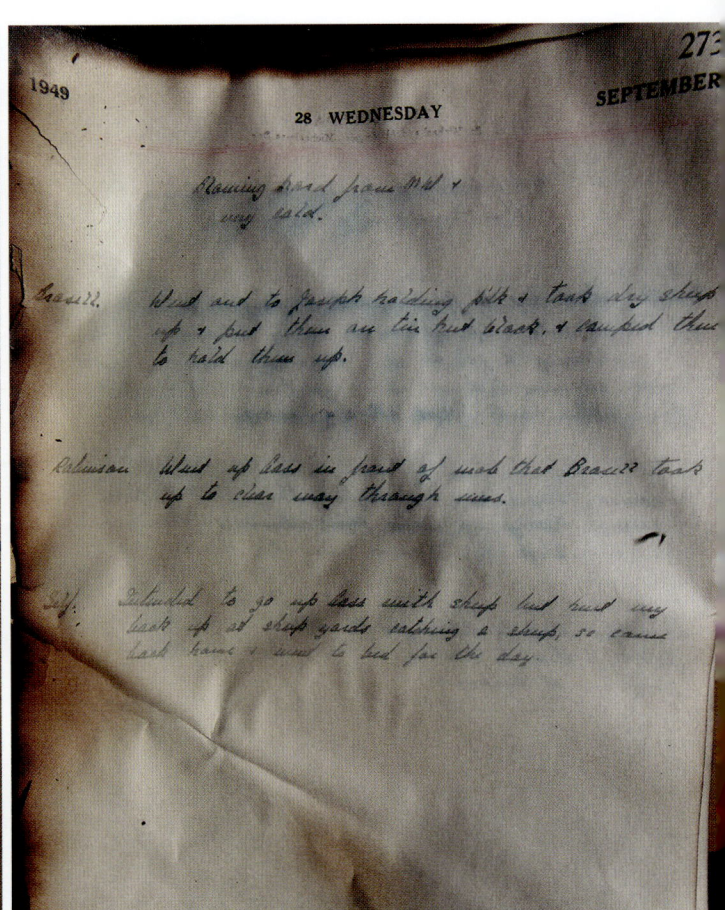

smell — I'll never forget it. The thing I found most gutting was losing my rifle — I loved that rifle. I can remember walking through the house, up to our knees in ash, looking for things. It was something you never forget, to see your parents that upset.'

The wider response to the Murrays' loss was overwhelming. Mackenzie lore emphasises individual resilience, pioneers who take on our harshest country to carve out an existence or perish. What's missing in that picture is the role of the Mackenzie Country community, which in bad times comes into its own.

The fire left the Murrays with only the clothes they were wearing. By the evening, they were established in the shearers' quarters. Scores of friends and neighbours arrived to help in any way they could, bringing bedding, toothbrushes, pens, loo paper, legs of ham — whatever was needed to get the family through the first few days. 'They were bloody amazing,' says Kate. 'We arrived in the shearers' quarters and the corners of the toilet paper were folded over and there were new toothbrushes laid out.'

Anne and Jim were overwhelmed by the support they received from the

ABOVE

The burnt remains of George's and Gerald's diaries.

RIGHT

The press report of the fire.

people of the Mackenzie. 'In an isolated rural community, everyone has a sense of belonging,' says Anne. 'Everybody came, however well we knew them. They arrived with all sorts of things from food, clothing and bedding to jars of buttons and Christmas decorations, all those mundane things that you tend to take for granted and that are of huge value to one's daily living.

'Both Jim and I were terribly worried about being able to remember who had given what so that we could thank them, so we appointed Erf, who was teaching in Tekapo, to act as quartermaster. The generosity from our friends and neighbours just went on and on. Parcels and letters were pouring in for weeks.'

(The Murrays themselves were known for helping out in a crisis. When Godley Peaks lost its woolshed roof in a huge blow on the eve of one shearing season, they offered the Scotts the use of the Glenmore shed, along with their shearing quarters. 'Jim and Anne were right behind us and helped right through that shearing,' remembers Liz Scott. 'We were hugely indebted to them.')

With so many people coming and going, Anne reverted to her role of high country hostess. 'I kept thinking, They've come to help, how do we feed them? While the fire was still smouldering I asked a couple of the men if they could get past the hot ashes to the big chest freezer which had been in the storeroom because I knew I had some hams in there. They got a jemmy bar and managed to wrench up the lid, but the contents of the freezer had been reduced to charcoal.'

When the ashes had cooled, people returned to help the Murrays sift through the debris. There wasn't much to salvage — a couple of crystal glasses, a tiny clay pot that had withstood the heat, Will's rowing medals and a bottle of boiled whisky. Everything else had been reduced to ash and cinders. Silver cutlery was warped and twisted; 105 years' worth of Jim's grandfather's and father's farm journals were charred beyond repair. A few rose bushes in

Tekapo homestead gutted

Fire gutted the Glenmore Station homestead (pictured at left) on the western shore of Lake Tekapo soon after noon yesterday. Two garaged vehicles were also destroyed.

The large, comparatively modern dwelling built by the owners, Mr and Mrs Jim Murray, in 1974, was ablaze from end to end when two units of the Lake Tekapo Fire Service arrived 20 minutes after the alarm was raised at 12.15pm.

Thirteen brigade members fought the fire for up to an hour but could do little except contain the outbreak. They were assisted by a helicopter, piloted by Mr Gavin Craig, using a monsoon bucket.

Water was drawn from a nearby swimming pool and a service tanker. Neighbours and bystanders pitched in and helped fire fighters with hose deliveries.

Among those helping was neighbour and Mackenzie District Council chairman Mr Bruce Scott, of Godley Peaks.

The Lake Tekapo chief fire officer, Mr Bill Apes, said the brigade made the trip up the Godley Peaks Road in 20 minutes. "But the place was burning from end to end and it was gutted.

"It took from about 45 minutes to an hour to bring the blaze under control," Mr Apes said. "We were only able to save a few personal effects from one room.

The large L-shaped, timber-framed dwelling with walls finished in natural greywacke boulders was unoccupied at the time. Mr and Mrs Murray and their family were busy mustering cattle up the Cass Valley.

Mr Apes said that only the outer walls remained.

"When we arrived most of the iron roof, except for one small portion, had caved in and flames were shooting 10 metres into the air."

When the fire had been brought under control the brigade spent the remainder of the afternoon damping down hot spots. During the callout, Fairlie fire appliances stood by in Lake Tekapo.

The alarm was given by a couple staying at a fishing bach on the 20,600ha station and Mrs Ann Marshall, wife of a station shepherd, rang Lake Tekapo for the fire service.

Mr and Mrs Murray were still too "shell shocked" to comment directly to a reporter last night but, through Mrs Marshall, indicated that they had been mustering cattle up the Cass Valley at the time. They thought the fire might have been caused by an electrical fault.

Two vehicles in an adjoining garage were lost in the blaze. The house and its contents were fully insured.

Mr Scott said "devastating" was the only word that could be used. A homestead was the pulse and engine room of any high country station.

Mr and Mrs Murray had worked so hard in the last 15 years to develop and enhance a superb and beautiful home. All thought it was indestructible.

"The phone has been red hot with offers of help but what can you do when this happens?"

ABOVE

Fire crews work desperately to put the fire out.

front of the house had survived and were later transplanted, so at least the Murrays could maintain some of their original plantings.

After his initial shock, Jim squared his shoulders and got on with things. 'I'd lost everything, lost our history, all those things that could never be replaced. But logic kicked in. In January Gordie McMaster came to visit and we were still pretty shaken. He said, "What are you worried about? You've still got each other, your family, and you still have your engine room."'

Says Anne: 'Jim will look at a problem like the 1967 snow or the fire and break it down. He is very practical: "So this is where we are at, and this is how we move forward." There's no panic, no sense of "How the hell can we do this?" Just huge determination and practical application.'

As for her own reaction, 'The first thing I thought was, we're all alive and all okay, but then I couldn't sleep for six weeks. I was in a kind of shock.'

The project to rebuild the homestead became a welcome distraction. Within days their helpful insurance company gave Anne and Jim the go-ahead to buy essentials — everything else would be processed in short order. It was the biggest domestic claim the company had ever paid, covering not only the homestead and contents, but also vehicles and farm equipment that had been in storage in the burned-out garage.

They had designed the homestead themselves, aged twenty-six. This

ABOVE

Will searching the ruins.

time, because of the urgency, they engaged an architect, but instructed him to take from the best of what they'd lost. The new house would have more space upstairs, with recessed dormers to let in additional light. But in many ways the new house mirrored the old. Jim was insistent on reprising the large verandah to keep out the weather and snow. He also made sure that it was well protected against fire, with a central sprinkler system and smoke alarms covering every room, and two fire-hose reels operating off a 5000-gallon tank.

Much of the furniture lost to the fire had been in Jim's family for generations. When the Murrays contacted Peter Hopkinson, the Temuka-based antique dealer who had valued the pieces for insurance, he suggested they join him on a scheduled trip to the UK to find replacements. With the build now well underway, Anne took him up on the idea.

After so much stress and heartbreak, it was a perfectly timed break. 'I had a ball,' says Anne. 'I was ringing Jim constantly for more funds. In the end I bought desks, kitchen tables, dining-room tables, crockery, cutlery, floor rugs, enough to fill a container. Some of it was old, some of it new. Peter Hopkinson was able to organise the customs clearances and to bring everything home.'

While she was away, Kate and Pip stepped into her shoes, feeding the seasonal team of crutchers. You'd expect that they'd have been naturals, but far from it, says Kate. 'We didn't know how to cook mince! We boiled it up in a pot and served it to the crutchers. Dad was just appalled.'

Ten months after the fire, the Murrays shifted into their new home. Anne's reaction wasn't what she'd expected.

'I burst into tears, saying, "I don't want to be in this new house." I was finally grieving for the old homestead. It wasn't until we moved into the new place, having had a happy year in the shearers' quarters, that I realised what we had lost. I was mourning the loss of the children's old Correspondence School gear, the bad-taste cushions, the hole through the door where the kids had fired their toys. It was a sense of lost history, and it affected both of us.'

They couldn't wallow. The fire had to be consigned to the past, filed alongside the snowstorms and the floods, while the focus returned to the challenge at hand. The farming operation hadn't broken stride; in fact, Glenmore was producing its best wool ever. But now a battle was looming, at Glenmore as elsewhere in the high country, for a greater say in the sale of its product. Merino growers such as the Murrays were about to fight for control of their future beyond the farm gate.

In the summer of 1994, Jim and Anne holidayed in the Marlborough Sounds, where they fished, swam and generally recharged their batteries after a big year at Glenmore. One afternoon, following some fishing in the outer Pelorus, they decided to head ashore at Pohuenui Island. It was a chance decision, a spur of the moment thing — and it turned out to be one of the smartest accidental moves the Murrays ever made.

Pohuenui is a private island of over 2400 hectares, historically the home of a merino farming operation which these days is supplemented by small-scale tourism. In 1994, the farm was managed by Brian Brackenridge, whose name would later become familiar as the co-founder of Icebreaker.

Jim: 'The fishing was no good, so I said, "Bugger it, let's go for a cruise in this bay here." We pulled into the wharf and this guy came trotting down and said, "Gidday, come up and have a cup of tea." It was Brian. We must have spent a couple of hours yakking away.'

The two men found they had plenty in common, not least a mutual sense of frustration at how New Zealand merino was being marketed and sold internationally. But that they should be in such close agreement wasn't surprising — anger over the industry status quo was as common as sunburn among merino farmers in 1994.

In *Dust to Gold: The inspiring story of Bendigo Station* (2009), Central

Otago farmer John Perriam describes the founding of breakaway organisation New Zealand Merino that year as a 'merino revolution', which imparts something of the flavour of hostility that had grown among his kin towards the old guard of the industry.

A leader among merino growers, Perriam explains how the New Zealand Wool Board, whose job it was to promote New Zealand wool internationally, had done next to nothing to make a case for merino. As a member of the International Wool Secretariat, the board had signed up to a generic wool marketing campaign aimed at reclaiming ground lost to the global synthetics industry. There was no room in such an approach to promote merino, let alone the special qualities of the New Zealand version, which Kiwi growers believed to be a superior product. Instead, New Zealand merino was thrown in with unknown merino fibres from around the world, with no identity past the first stage of processing. 'We weren't being rewarded for having a better product,' writes Perriam.

The obvious answer was to ring-fence the local wool. The more enlightened merino farmers also realised that the back-story of their product, grown among the snowy peaks of high country New Zealand, made for a compelling marketing tool. Perriam's experience of Japanese and Italian buyers was that they would pay a premium to be able to use that story.

Like any coup, this one began with the plotters working the phones in an attempt to get the numbers. Jim Murray was among the ringleaders, and eventually took a place on the founding board of New Zealand Merino.

'We were pissed off with merino wool being handled by the Wool Board as a commodity product, paying huge levies to that board and going absolutely nowhere,' he says. 'Until New Zealand Merino got cracking, the world thought that only Australia and South Africa produced merino wool. And even if we contributed only two per cent of global merino production, our research showed that we were in the very top echelon in terms of quality. That spoke for itself, really. Something had to change.'

At the time, the Wool Board collected a compulsory levy of six per cent of value per kilogram of all wool sold, which was used to promote the Woolmark brand. The merino lobby demanded that it be handed over. Perriam: 'What followed was outright warfare and dirty politics. The Wool Board fought with every means possible.'

After months of wrangling, the board grudgingly agreed to hand over the funds, on the proviso that the merino growers conceded to being an advisory committee only. Cue a resumption of hostilities and, following the intercession of the Minister of Agriculture on behalf of the rebels, the birth of New Zealand

ABOVE

Evening light on Glenmore's homestead paddocks as sheep wait to be fed out.

Merino, a 100 per cent grower-based, merino-specific marketing body. Perriam became chairman on the death of founding chairman Robert Jopp, and Brian Brackenridge's brother John was appointed CEO.

Yet it still wasn't entirely independent of the Wellington-based Wool Board, even after setting up a separate office in Christchurch. 'We were a lightweight organisation taking on the heavies,' says Jim. 'We believed we had a right to that levy, and for it to be totally market-specific to fine wools. There was a lot more to-ing and fro-ing, but we got there in the end. And then we were away.'

During the next five years, the compulsory levy was finally able to be used to brand New Zealand merino on the world stage. New Zealand Merino formed relationships with luxury international garment-makers such as John Smedley and Loro Piana, and the marketing team kept winning awards. (In 2000, a consultant's report instigated the transformation of New Zealand Merino into a commercial company with a sales and marketing focus, the New Zealand Merino Company Ltd.)

A 2009 University of Canterbury study on the marketing of New Zealand merino shows just how radically things changed in the years after 1994. Where previously almost all of the New Zealand merino clip was hawked as a commodity product at auction, by 2009 something like eighty-five per cent was going out under the New Zealand Merino brand.

'It was a euphoric time for us,' recalls Jim. 'Finally, New Zealand had become identified in its own right as a producer of high-quality merino wool.'

By choosing to contract to supply a particular company in New Zealand or overseas, the merino growers also got security of income. 'By and large at Glenmore we've done better under contract than if we'd been in the auction system,' says Jim, who describes the creation of New Zealand Merino Inc. as a 'milestone' in the history of the breed in New Zealand.

'During those early days I gave a talk in Australia, where they were still very much locked into the Australian Wool Corp. I said we're small, we're mobile and can shift direction very quickly, and the relationships are more one-on-one. This was fifteen years ago and the Australians then were very envious of us, to have such a manoeuvrable and motivated organisation marketing our merino to the world.'

While all this change was in the wind, Jim had been back to see Brian Brackenridge, intrigued by something he and Anne had been shown on their visit to the island. It was a T-shirt made of fine merino wool, incredibly soft and wonderfully silky. Brian and his wife Fiona were planning to manufacture and market a range of these garments, and they already had a name in mind: Icebreakers.

Says Anne: 'We looked at these shirts and thought, wow.'

Jim: 'By the time we got back to the bach at Mahau Sound I said to Anne, "This thing has got to be a goer."'

Unbeknown to them, in the weeks between their two meetings with Brian, a twenty-four-year-old marketing graduate from Wellington had also met Brackenridge. 'We were thinking about what to do next, tossing around ideas, and blow me down if Jeremy Moon didn't come in the other door,' says Jim with a laugh. He later rang the young entrepreneur, who committed to buying wool from Glenmore to make the very first Icebreaker underlayer garments.

By the end of the year, Jim and Anne had made a substantial investment in Icebreaker. 'It was all about the product,' says Jim. 'There was nothing else like it on the market.' His first impressions of Jeremy Moon? 'Who the hell is this scruffy bugger? But I always enjoy someone who is motivated, focused and prepared to give things a go. We didn't know if he was going to be successful, but he had some brilliant skills at marketing.'

For his part, Moon found in Jim someone who could teach him not only about the attributes of merino wool, but also the essence of high country station life. 'The unique thing about Jim was that he was a wool-classer as well as running Glenmore,' he remarks. 'I was trying to find out what attributes of the fibre made a difference to fabric quality. I found out it was all about the micron and tensile strength, and the co-efficient of variation of fibre diameter and length, all of which Jim taught me, rather than what a lot of the classers and growers prided themselves on, which was this mysterious idea of "style". So we were destroying a few myths in the process.'

'And Jim's an innovator,' he continues. 'He is traditional on one side, but

ABOVE

The Run 79 Merino Gallery at Lake Tekapo.

RIGHT

'Soft-nosed'.

he is very interested in new ideas.' Example: supply contracts. 'At the time, the idea of contracts went against the grower mentality, which was more of a gambler's mentality, gambling on the market,' says Moon. 'We were trying to create a win–win, where the manufacturer in our case, and the grower in the case of Jim, Anne and Will, built a mutually beneficial long-term partnership to ride through the highs and lows of merino production.'

Moon's other great teacher was Glenmore. From his very first visit, when he felt himself 'enveloped' by the dramatic setting, the station set the tone for his fledgling brand. 'It was the towering mountains, the incredible scenery with the lake, the size and grandeur of the station. I remember having the experience of almost being able to feel the pulse of the place. I found it quite a spiritual place that connected with me very deeply. There was an incredible combination of animals, place and people, and those relationships became a foundation of the Icebreaker brand.'

As exclusive supplier during those first few years, Glenmore was Icebreaker's 'shop front', its romantic provenance story. Jeremy Moon brought several parties of overseas retailers through Glenmore to experience the farm in operation.

'We had Chinese, Indians, North Americans, English, all different groups here,' says Jim. 'One time I got our neighbour, the late Bruce Scott, to head up an advance party to Waterfall Hut and produce some camp-oven scones and billy tea with that wood-smoked taste. These guys thought that was absolutely wonderful — and here they were looking around at the land from which this beautiful merino wool came. It was a fun time getting to know those overseas people, because the merino industry was really on a roll.'

But Jim and Anne weren't satisfied with just supplying wool. 'The dream was always to take the Glenmore wools from the sheep's back right through to retail,' says Jim. 'Soon afterwards an opportunity arose to establish an outlet in Tekapo.'

Run 79, as they named it, would have exclusive rights to retail the Icebreaker brand outside the main centres, along with other merino garments. It was a big step outside their comfort zone, but that was part of the excitement. In any case, establishing an alternative, non-farming revenue stream was long overdue.

They found dress-circle premises in Tekapo village, then Anne disappeared to Christchurch for a crash course in retail presentation, filching ideas from some of the city's smarter boutiques. When it opened in 1997, the Run 79 shop décor played up the origins of the merino garments, with a large map showing the Mackenzie Country pastoral runs, a handsome display of fleeces and items from Glenmore's shearing shed.

On opening the doors, they crashed head-on into a perception problem. Consumers today are aware that merino has unique qualities as a natural temperature regulator, and that it is kind on both skin and body odour. In the mid-1990s, however, people weren't sure what to think of it.

'It was tough at first,' says Anne. 'There was a huge perception of merino

ABOVE

Will on Tin Hut Block after an early fall of snow.

being prone to shrinking, pilling, itching, and so on. It took a lot of education and usage to dispel those myths. But once somebody purchased a merino garment, they tended to become regular buyers. The proof was in the wearing.'

Some of Run 79's most enthusiastic customers were international tourists, who were impressed not only by the quality of the garments but also by the romance of the story. Spying an opening, the Murrays launched into online retail. Their timing couldn't have been better: Icebreaker wasn't yet being sold overseas, so Run 79 was able to capture a significant international market.

'About half our time was spent doing overseas orders — more often than not, repeat orders. The business was hugely successful,' she says. In fact, Anne became so busy she had to employ a shop manager, Lynne Frost, who in 2009 bought the business from the Murrays.

A few years earlier, they'd sold their Icebreaker stake. 'An offer went out to all the non-family shareholders. It was appropriate at the time [to sell],' explains Jim. 'But I've always said that when we set up New Zealand Merino we missed our putt. We should have bought Icebreaker at the same time and retained Jeremy as the head. I could see how good it would have been for the merino industry to have it.'

The decade following the mid-1990s merino revolution had been a heady time, and Glenmore did better than most to capitalise. But the industry today is a very different beast, according to Jim. Prices have plateaued, while costs just won't stop climbing. 'It's a big worry.'

Jim frets, too, that the contemporary version of New Zealand Merino is chasing too many issues. He applauds the efforts to reposition merino as dual-purpose sheep, good for both wool and meat. But too many peripheral issues are clouding the organisation's focus. 'New Zealand Merino has taken its eyes off the ball. It's still got some fantastic opportunities, but it needs to get back on track,' he comments. 'The processors, the retailers and all the different merino brands have this big challenge to meet in the global move towards more casual clothing, which is dampening demand for fine wools.'

Will and Ems, meanwhile, have signed up to long-term 'whole clip' contracts to supply the New Zealand Merino Company, with Icebreaker as the end customer. Glenmore's micron, says Will, is very well suited to Icebreaker's contractors. 'It's right on what they want for their sports apparel and, because we've got our own stud and breed our own rams, our micron doesn't fluctuate a lot. That makes it easy for the New Zealand Merino Company to give us a whole-clip contract because they know what the micron measurement is going to be for the next two or three years.'

There's a risk in signing long-term contracts, concedes Will. 'Auction prices can always go higher. But that's only happened once in the past decade. And it's nice to have that ongoing association with Icebreaker. We get a lot of feedback from the company about our wool, too, such as what wool performs best, and how it should be classed. That's important to us. Too often in the past you'd sell your wool and once it had left the farm gate you had no idea where it went. Now we know exactly where it's going. If we want to, we can click on a barcode and bring up a video of Glenmore.'

Yet even with the security of fixed contracts, Will and Ems can't relax. Prices for merino wool are in the doldrums — have been for some time. Cheap merino products coming out of China are a factor. Among Glenmore's neighbours, Balmoral Station saw the writing on the wall and recently launched its own merino garment brand. Others are diversifying into property and forestry.

'We're still getting ten dollars per kilogram for our wool,' says Will. 'It's stagnant, but at the same time our costs keep climbing. Your only option to increase your wool cheque is to produce more of it.'

The other approach, however, is to reduce your reliance on that wool cheque by finding new income streams. As will be seen, that's exactly what Will and Ems are endeavouring to achieve.

Fishing huts at Lake Alexandrina.

CHAPTER FOURTEEN
LAKE ALEXANDRINA

Lake Alexandrina, the narrow waterway that constitutes Glenmore's southern boundary, is so beloved by fishermen that some truly can't bear to leave the place, even in death.

Dr Christmas, late GP of Timaru, is buried on the plateau at the western end — or at least his ashes are. George Barber is on the eastern side, between two Oregons he planted in one of those rare moments when he wasn't catching trout, filleting trout or eating trout. The accountant A Jones, also of Timaru, is scattered elsewhere on the lake shore.

LEFT

A fishing hut at Lake Alexandrina.

Of Alexandrina's three fishing camps, Top End, which dates back to 1918, is the smallest by far, just fourteen huts of varying degrees of quaintness perched on Glenmore land between homestead and lake. The owners have no title. They pay rates, collected by Will, who is billed in turn by the district council, but there's no guarantee of access.

Since Gerald's day, however, it's always been understood that the Murrays are happy to have them. Symptomatic of the relationship between the farmers and the fishermen, one hut at Top End is constructed entirely of timber and iron leftovers from Gerald's woolshed. These baches are fiercely protected family assets, with ownership passed down through the generations.

What's the appeal of Alexandrina? It's beautiful, certainly, a spring-fed oasis in a bare, dry country, its waters a sharper blue than Tekapo, where the suspended glacial 'flour' produces a milky turquoise glow. Viewed from Glenmore's high downs on a windless autumn morning, Alexandrina is a bright mirror to the Mackenzie sky.

During particularly cold winters, the northern end of the lake can freeze solid. 'My father once lost thirty beef cows in Lake Alexandrina,' recalls Jim. 'The lake had iced over in part and it was covered with a light layer of snow, and the cows had walked out towards the middle, not being able to tell the difference between ice and hard ground beneath their feet. The ice broke, they fell through, and were discovered floundering in the freezing water. It would have been suicide to try to retrieve them, and they all perished. At that time Gerald only had sixty to sixty-five cows in his entire herd, so it was a big loss.'

Bird life is another attraction. During summer, marsh crakes breed around the reed beds, before migrating to coastal parts of Canterbury. Alexandrina has the largest population of southern crested grebes, and is home to good numbers of New Zealand scaup, the little black diving duck with the golden eyes, as well as paradise ducks, Australian shovelers and many others. Ornithologist Ray Pierce describes Alexandrina as a 'critically important' waterway to several of the forty-five-odd species found there.

Ray has a soft spot for the quiet Mackenzie lake. 'When I think of New Zealand, I think back to the years when as a kid I camped at Alexandrina. The lighting there is just so special, the colours of the hills changing with each hour of the day.'

Yet other lakes have beauty and bird life, too — the fishing is what makes Alexandrina special. It's not so much the abundance of trout as their size that draws anglers from all parts of New Zealand.

Roger Shaw has loved Alexandrina since he was nine years old. On the Shaw family's first visit to the lake, Gerald Murray towed their 1936 Vauxhall and caravan behind three draught horses through what was then undeveloped swampy country to Top End. At the end of the holidays, Roger was sent up to the homestead to ask Mr Murray to tow them out again.

Seventy years later, Roger still regularly makes the trip from Temuka to the family's bach, an army hut that his father had trucked in from Temuka in 1945. 'Mine is one of the roughest at Top End,' says the retired carpenter. 'We have a sixteen-foot caravan under a port there, with an eight-foot-square room on the back and an outside toilet. It's what I call a good Kiwi bach.'

When he was younger, he used to fish sometimes until four in the morning. Even now, it's often past midnight before he turns the rowboat for home. 'I like the changing of the light at night,' he says.

There are many more anglers fishing today, and the fishing has got tougher. But those monster fish are still there — growth rates at Alexandrina are among the very fastest of lake fisheries in New Zealand.

RIGHT

Reflections on Lake Murray.

'Three years ago I caught a brown trout in Alexandrina that was thirteen pounds. My wife got a ten-and-a-half pounder off the bank at Top End, and I've caught two ten-and-a-half pound fish, one a brown, the other a rainbow. I'd sooner go there to try to catch a fish than the Opihi River, which is virtually at my back door at Temuka.'

It was John McGregor, the early runholder of Glenmore, who in 1881 liberated the first trout in the lake. (Alexandrina is named after a sister of McGregor's farming partners at Glenmore, John and William Robinson; a smaller adjacent lake, into which Alexandrina drains and which in turn empties into Tekapo, is named for the Scotsman himself.)

In *High Endeavour*, Vance describes the pains taken to bring those fish to the Mackenzie. 'The young trout were sent from Christchurch by train to Albury and taken by coach from there to Burkes Pass by Elijah Smart and Maxwell Black. It was midnight when they arrived at Burkes Pass, where McGregor was waiting. He took possession and drove straight-away to Lake Alexandrina and liberated them there just before dawn. All the trout, except three, survived the long and varied trip.'

From 1923 until 1977, the South Canterbury Acclimatisation Society (SCAS) pursued a long-term policy of liberating more fish into Alexandrina — almost seven-and-a-half million of them, in fact, the majority rainbow trout. Such large-scale release was required because of the limited usefulness as spawning grounds of the two main tributaries to Alexandrina, Scotts and Muddy creeks. Crowding of the Scott's Creek spawning ground in particular may partly explain why the fish in Alexandrina become so big so fast.

The society has historically kept a close, sometimes anxious, eye on the lake. A report on an SCAS meeting in 1917 noted worrying accounts of trout dying in their hundreds. 'On investigation it was found there was only a small percentage, and there was no cause for alarm,' it soothed.

In the 1980s, the appearance of algal blooms on the lake stoked tension between the society and the Murrays, culminating in an attempt at a moratorium on farming in the Alexandrina catchment. The headline in the *Timaru Herald* the next day read 'Glenmore Farming Halt Called For'.

A study by Lincoln scientists had concluded the level of nutrient build-up in the lake was unusually high. The finger was pointed at Jim's agricultural development of the surrounding country. Were they right?

Jim argued that the science was thin: there'd been no baseline data recorded until after 1980, so no proof of a deterioration. As well, there was no clear understanding of how Alexandrina was supplied; along with those tributaries, some of its water comes from underground springs fed by the

Cass. He noted his stocking units were actually lower than a decade earlier, and that he had abided by superphosphate levels agreed to by the society. Finally, he revealed that his father had observed algal bloom back in the 1930s, when Glenmore had been a low-input pastoral run.

A counter-argument was that a causal connection between eutrophication of waterways and agricultural activity was already well established. If Glenmore wasn't responsible for the unsightly blooms, which at times blanketed the littoral area of Alexandrina's northern end in blue-green scum, then what was? Investigations showed minimal contribution from the fishing camps.

The push for a moratorium failed, but Jim volunteered to fence off the two creeks and created a buffer zone of sixty acres at the head of the lake. He also agreed to cultivate no more than one paddock at a time in the Alexandrina catchment, and to monitor lake water quality for any deterioration.

Will and Ems are continuing with the water-quality monitoring programme. Defined by ECAN's new Canterbury Land and Water Regional Plan as a 'Sensitive Lake Catchment', there are plenty of restrictions on farming activities in the catchment, including limits on how much nitrogen can be leached, a ban on ploughing up, and a stipulation that only one paddock can be renewed per year. 'Any fertiliser is put on by truck and kept well away from waterways,' adds Will. The new spray irrigation, meanwhile, is destined for a catchment that drains into Tekapo rather than Alexandrina. Already, Will and Ems have begun intensive water-quality sampling of Mailbox Creek to establish a baseline, and have fenced off all of the creeks that drain into the larger lake.

'The water quality generally isn't bad at the moment,' says Will, who has never seen an algal bloom on Alexandrina. 'It's a beautiful lake,' he adds, 'but we don't have too much to do with it. It's not a great swimming lake and you're not allowed jetboats on it. The most fun we have as a family with Alex is actually during the wintertime, when the top freezes over from the island all the way down to our end. At night when it's cold there's this really eerie sound of the ice cracking, like a rifle crack shooting straight across the lake. We tell the kids it's the sound of the Alex ghost.'

There are more than enough fishermen whose ashes are scattered at Alexandrina to raise a posse of ghosts — although Roger Shaw has never seen a single shade. Approaching his eighties, he'll keep chasing the trout while he can, he says. Will his own ashes some day be scattered at Alexandrina? 'I like the place — in fact, I love it. But being buried there? No, I'll be in the local cemetery.'

Mustering out of Tin Hut Creek.

CHAPTER FIFTEEN
PREPARING THE WAY

Living in the eye of the tornado, Anne calls it. She's referring to the closing period of their tenure at Glenmore, when she and Jim became online retailers, property developers and second-farm owners all in the space of a few years. Or perhaps she's simply talking about living with Jim, whose entrepreneurial energy was never going to be happily contained within the boundaries of a high country sheep station.

By the 1990s, the early pell-mell development of Glenmore had quietened into consolidation mode and Jim was antsy. He needed a new bone to chew on. As well, with a succession plan for Will to take over Glenmore now firmly established, they needed to find new sources of income. Not only did he and Anne want to firm up their retirement funding, there were Kate and Pip to think about, too.

'We focused on creating off-farm assets so that we could be in a reasonable position to help the two girls and Will into whatever venture they wished to pursue,' says Jim.

Their first move, however, was to buy another farm. 'We'd had a couple of years of very good wool returns. I'd paid off the mortgage with the deer, we were debt-free and looking for something to invest in. All the professional advice was that I should stick to my knitting — farming.'

Lynton was a 648-hectare dryland farm at Hororata, at the mouth of the Rakaia Gorge, with light and stony soils that struck Jim as perfectly suited for running merinos. Not only that, but the previous owner was Peter Smail, a highly respected figure in farm forestry — in Canterbury he was known as the Man of the Trees. Every paddock at Lynton was surrounded by trees, guaranteeing shelter from the scorching Canterbury winds.

'It was a beautiful property. At the time the merino industry was on a run and wool prices were excellent — and I thought I knew a little about them — so we built it up with merinos from Glenmore.'

Lynton was a far more modest proposition than Glenmore, but there was no way Jim could run both, although he visited at least once a month and every spring classed wool at shearing time. Chris Ensor, who had worked at Glenmore for several years, was transferred to manage Lynton as a stand-alone unit, running 6000 merinos. 'It was operated entirely separately and had to stand on its own feet,' says Jim. 'We stocked it from Glenmore, but Lynton had to buy that stock. Everything had to stack up in its own right.'

A decade later they sold up. Jim says he never regretted the decision to buy Lynton, but it was time to shake free from the vicissitudes of the commodity price cycle. Cashed-up and looking for an off-farm investment, he and Anne leaped when an opportunity arose to buy a commercial development in Tekapo village. The block included the Tekapo Hotel, motel and nine retail outlets, including Run 79. They became landlords.

'I didn't have a bloody clue what I was doing but it was brilliant,' says Jim. 'I got an insight into commercial management of buildings, leases, tenants and everything else that goes with an operation like that.'

Indeed, he was enjoying it so much that when the district council came

RIGHT

Jim out grading on the Cass River accompanied by his dog Ouf.

knocking with a proposition to develop a new commercial precinct, Jim was tempted. For sale was eight acres of council-owned prime real estate on the lake shore. The local authority wanted to see it developed and thought Jim should buy it, build the infrastructure and buildings, and then lease it out.

'The setting was fantastic. The development was exactly what Tekapo needed. But I could see that it would be commercial suicide.' They passed, and when someone came along offering to double their money on their existing complex, Jim and Anne didn't hesitate.

While all this deal-making was taking place, Will had returned to Glenmore, and Pip and Hamish had married and were looking to go farming. The succession clock was ticking. 'Anne and I weren't far away from moving off the property, and I needed something to keep me out of jail,' says Jim of the final stage of their slow, methodical self-extraction from Glenmore. Cue property development.

Lochinver Run was once the home block of Tekapo House, a nineteenth-century hostel on the edge of the Mackenzie Plains. (It was a requirement of the day that all rural hotels must be able to graze animals to feed the travelling public.) Today, the land is a residential subdivision in the last stage of a three-stage development, the final forty-eight lots being marketed with a pitch to 'Own your piece of the Mackenzie Country'.

Apart from a long break following the global financial crisis of the late 2000s, the development of Lochinver has occupied Jim for the past few years — kept him out of jail.

'I was on a hell of a learning curve again,' he says of this foray into property development, achieved in partnership with Anne and their friends John and Marion Murphy from Wanaka. 'One had to come to grips with the process of consents, dealing with design and engineering firms, with council and all the rest of it.'

His experience running Glenmore proved helpful to a degree. 'It was the advent of the consent compliance regime. We'd seen it, we were ready for it, although this was a totally different arena. There was plenty of expertise out there, and we certainly tapped that.

'We programmed three stages. Stage one was forty-five sections. We planned it, had the sections pegged out and produced a map, and then put the sections out for individual tender. We had two hundred and twenty tenders come in for those forty-five sections! Obviously, there was a pent-up demand for land in Tekapo. Stage two was twenty-two sections and they also sold well — all but a couple went straight away. We hit a brick wall in 2008, but it didn't worry us; we were in a financially secure position and so we rode it out.

ABOVE

Joseph Hut, formerly the boundary keeper's hut.

We're back into doing stage three now.'

While Lochinver was going down like a smooth single malt, Jim and Anne were struggling to pull off a second property project — this time at Glenmore. The plan was to subdivide four hectares of their freehold lake-front land. They had a buyer lined up, an Aucklander whose plan, says Anne, was for a house that was 'low profile, of non-reflective and natural materials, below the skyline and that complied with the district plan'. But there were objectors, among them Forest and Bird, the Aoraki Conservation Board, the local community board and ECAN.

The latter two organisations were mollified after changes to the plans, but Forest and Bird and the Aoraki Conservation Board weren't persuaded. The subsequent hearing before a commissioner cost $120,000, and, after the Murrays had been given the tick, the Auckland buyer walked away.

It ended well enough: the land and plans were sold, the proceeds used to help Kate into a house in Wanaka. But for Anne, the process confirmed the

Ems and Will mustering cattle for weaning in the Joseph paddocks off the Cass flats.

intransigence of some lobbyists in the Mackenzie. 'We said, "Look, you won't even see this house," and we explained what it was about, and that we had met all the requirements of the district plan, and to look at the big picture of us trying to help our family, but . . .'

In 2003, Will and Ems began farming Glenmore in a fifty–fifty partnership with Jim and Anne. Three years later the younger couple bought the balance of stock and plant, signalling the start of Will's time at the helm.

With the equity from the sale, Jim and Anne built a new home for themselves at Lochinver, allowing Will and Ems to shift from the cottage into the homestead.

For Jim, Tekapo was a fair fit. 'I could trot off to Glenmore and be totally occupied,' he says, 'but there was no infrastructure in Tekapo for Anne, nothing to occupy her mind. I can understand that there wasn't enough in Tekapo to keep her going. I said to Anne, "You've spent forty-three years of your life with me at Glenmore, so it's your call where we go next." We had three years at Tekapo, in a beautiful home with a wonderful outlook, but the choice we made together was that we move either to Wanaka or to Timaru.'

BELOW

Angus and Jim — tailing time.

Anne suggests the move to Wanaka was also about giving Will room to make his mark at Glenmore. 'It was Will's time,' she says, adding that Jim is still very much involved in Glenmore life, just less frequently. 'Jim's done fifty-three or so autumn musters and at seventy he's still fit enough to keep mustering. He comes back and classes the wool at shearing time, and when they're classing up the stock for breeding, and at various other times of the year.'

But leaving Tekapo also gave Jim a chance to make a more definitive break with his old life, to shuck off his role as runholder of Glenmore Station. It was the moment to take stock of what he and Anne had achieved there.

'The greatest satisfaction for us was to leave Glenmore, the business side of it, in great shape for Will,' says Jim. 'We had farmed in an environment that was very challenging, but we felt confident we were leaving the property in good shape. We'd insulated ourselves against commodity price fluctuations by establishing diverse farm income streams, and I've always felt tremendous pride that in spite of the ravages of rabbits, hieracium infestation and dry seasons, the health of the land was still in a wonderful state, as were the buildings and fences we left for Will. I'm not criticising the way my father ran the place — it's what he wanted to do and I respect that — but what Anne and I did there was totally different, and we have a huge sense of satisfaction and pride looking around the place and the production coming off it. It was also gratifying that we had managed to help Kate and Pip with their ventures.'

It wasn't quite the final article, however; as Jim says, 'farming is fluid', and nothing is ever finished in the high country. Will and Ems would find they had plenty of challenges running Glenmore in a rapidly changing farming environment.

Will and Ems met soon after Will had returned to work on Glenmore under Jim. He'd completed his diploma in farm management — a 'dippy' — at Lincoln in the early 1990s, a time he remembers as being as much about learning to drink and have fun as studying the art of farming.

'I did a lot of my mountain climbing while at Lincoln. We stayed in the halls of residence and nearly got kicked out after six months. There were four of us rooming side by side, with a fair bit of partying going on, and eventually we were told by the vice-chancellor that it would be a good idea to leave before we got thrown out. We ended up flatting together in Christchurch the second year and had a ball.'

Post graduation, Gordie McMaster organised a year's work for Will in Australia, following which he travelled through Asia and Africa. He returned home on 19 August 1997.

'I always remember that date. When I came over Burkes Pass and into the Mackenzie it struck me again how beautiful the place was, how unique and how special it was to be surrounded by snow-covered mountains and that huge expanse of tussock land that gives you such an amazing sense of space.'

It's a common sentiment among the people who inhabit this rugged part of New Zealand, a sense of connection mingled with awe. Enough tears have been shed at the summit of Burkes Pass over the years to have irrigated the basin. As a boy, Will loved living in the Mackenzie for the opportunities it afforded to hunt tahr and fish for trout. 'But as I got older I came to appreciate it for other reasons. Coming over the pass I felt a real pride and sense of belonging, not just to Glenmore, but to the Mackenzie.'

He was back to stay, moving into the position held by Glenmore's married man, who went off to manage Lynton. Getting reacquainted with Jim's style as a boss wasn't always comfortable, he remarks. 'He expected things done his way, because that's just the way he is. But if you were prepared to work hard and use your brain and not muck around, you got on with Dad. Jim's a very strong man and, although I didn't agree with everything he said or did, we usually got on well.'

In hindsight, he can see that Jim had his reasons for putting him to the test. 'Coming back from overseas at nearly twenty-five, having spent the previous few years having a fantastic time, my feet weren't necessarily on the ground. Jim realised this. He thought I needed to knuckle down and get on with life, and I respected that. In the end I must have passed muster, because he started to go away for the winter with Mum, leaving me in charge of Glenmore.'

One cold evening while watching television in the cottage, Will heard the sound of horses clopping up the driveway. When he went out to investigate he found two young women riding bareback under a bright silver moon. Emily Rhodes and a friend had arranged with Jim to stay in the shearers' quarters for a trekking weekend.

Jim had known Ems's father Rob Rhodes at prep school, and each had been a groomsman at the other's wedding, but in the intervening years they'd lost touch. Their children had never met. For Will, who as a young man was shy of women, it was almost love at first sight.

Ems was working in the Timaru office of Sport Canterbury at the time, and training hard for the Coast to Coast. Searching for a reason to stay in touch, Will decided to enter the race, too. 'He wanted to have an excuse to ring me for training ideas,' says Ems. 'There was a phone call about how to fill in the application form, a phone call about how many times he should be training. A week later he rang me and asked me out for dinner. We met in Fairlie, and that was the start of a very intense courtship.'

The training continued, however. 'I was always saying, "You've got to be doing more,"' remarks Ems, who studied sport as well as commerce at university. 'What I didn't understand was how much physical work he was doing farming. He finished in the top quarter of the field.'

Maintaining a long-distance relationship is always tricky, but the Mackenzie threw in some special challenges. Will recalls a night when he and his good mate Hamish Smith from neighbouring Godley Peaks — now Pip Murray's husband — were returning from a game of squash in Tekapo. 'We were driving home through snow at ten o'clock and found Ems stranded on the side of the road. She was sneaking up midweek to surprise me, and in the snow she'd driven her car off the road and into a dry lagoon.'

It wasn't long afterwards that Will decided to ask her to marry him. 'Ems was coming on the Saturday, so on Friday I headed up to the Tin Hut with everything you would need for a picnic — a table, chairs, gas bottles and rings, and two bottles of wine. I carried it all up the hill and plonked it underneath an overhanging rock, and put the wine in the creek to cool.

'The next day it was raining, but I told Ems I needed to head up to the hut to take some photos. Top of the hill I gave her a flower I'd picked, got down on one knee and asked her to marry me. She didn't need much persuading. Then we went around the side of the hill and had a three-course dinner sitting in the drizzle among the tussocks and the tahr.'

They decided she should come to live and work at Glenmore right away. Ems had grown up on a South Canterbury farm, one of four Rhodes daughters who'd been expected to work as hard around the property as if they'd been boys. By the age of eight she was driving a tractor and doing stock work in the evenings and weekends.

'My upbringing gave me the ability to be a farmer in my own right and I had a love of farming, so I was fortunate to meet Will,' she remarks. 'We had a discussion about what I was going to do here, and approached Jim with the idea that maybe I could get some dogs and come work for him. He ummed and ahhed a bit, then said, "Alright, you won't be paid much, but you can come up."'

Moving to Glenmore was an eye-opener. 'I don't think people who live down-country have any concept of it, and that's true even of my own family. It's the length of the working days and the sheer physicality of living here.

Everything is physically hard work. In the wintertime with the snow it can be a battle just getting the kids to school; that can take hours, or you decide you're not going to try that day because the road's too icy or snowy and stay home. I don't think you grasp that until you have lived here for some time.'

The farming, too, was tougher than she'd imagined. 'I was on an apprenticeship under Jim, so a very high bar was set right from the start. He expected you to work hard, and if it took twelve hours each day to do what you had to do, you did it, no questions asked.

'The other thing is that I was starting from scratch with merinos. I had good stock sense, but merinos are so different from crossbreeds, which is what I was used to. Mustering on the hill was quite a learning curve, too,' says Ems, who managed to fit in five autumn musters before she had Angus.

'I remember saying to Will one night, "Everyone thinks living in the high country is this romantic thing, but it's just bloody hard work."'

Yet the place 'gets into you', says Ems, and even now after living in the Mackenzie for close to fifteen years her breath can still be taken away by the country and its moods.

They married in late 2002, and continued working the station under Jim's direction for a further three years. The handover in 2006 was a milestone, a new chapter. For Jim it was a kind of amputation. Glenmore was his birthplace. As a child it had been his entire world, as a young man it had been a proving ground, and then it became the place where he and Anne had raised a family. It wasn't going to be easy to leave.

RIGHT

Ems mustering cattle.

ABOVE

The team — always waiting.

Succession in farming can devastate families. Often it just becomes too hard to find a route through the economic and legal complexities, to answer questions of fairness among siblings and between generations. The farm will be sold, the connection to a historic property ended with a pen stroke. Look around the high country today and you find very few runs still held by the same family after multiple generations.

The Murrays, who are into their fifth generation at Glenmore, have bucked that pattern. Leaving aside for a moment the emotions stirred by the transfer, theirs has been an amicable transition, the result of sound planning, excellent communication and good intentions.

The land itself remains for now in a family trust. Will and Ems lease the property and buildings from the trust, but have bought the stock and plant.

Pip and Hamish have been helped onto own Ben Dhu at Omarama, Kate into a house in Wanaka.

Jim says that succession is particularly tricky when you have one child who wants to farm. 'One of the hard things about farming is that all your development work is not necessarily done for yourself, because the results of that work don't come on-stream for ten or twenty years, and one of the real issues is the tremendous capital asset that is tied up in the land. That's just the way land values have gone generally. So it's not possible to be equal, but you can be fair.

'From when the children were of an age that they could understand what was happening around them, Anne and I did our damnedest to give them the opportunity to have input so that each of them had a fair understanding of what was taking place in the bigger family picture. We have been very, very lucky to have the kind of relationship with our children where we can discuss things quite freely and openly.'

Talk of luck underplays the amount of planning involved. Most of Jim and Anne's off-farm ventures were undertaken with an eye on succession. 'We focused on creating off-farm assets so that we could be in a reasonable position to help the two girls and Will into whatever venture they wished to pursue,' says Jim. 'What a lot of rural families don't realise is that it probably takes between twenty and twenty-five years of planning to make it work.'

For their part, Jim and Anne have swapped the big skies and near-silent tussock land of Glenmore for Wanaka, a busy tourist destination that is rapidly evolving into a kind of 'Queenstown lite'. They've managed to find a property at the peak of a new hillside subdivision away from the bustle of the village, set among kanuka and tussocks and with mountain views. It's a beautiful location, just two hours down the line from Glenmore, but it's a world away from the high country, and for Jim it's not yet truly home.

Anne has watched Jim struggling with the transition. They ski, walk up Mt Iron every morning, and spend long periods at their bach in South Westland near Jackson Bay, where Jim likes to fish and go whitebaiting. Anne is involved with music and art groups, Jim is still neck-deep in the tenure review process — to his frustration at times. They still have property development interests at Tekapo.

But, as Jim concedes, you don't walk away from a place such as Glenmore and just flick a switch. 'His heart is in the Mackenzie Country,' says Anne. 'He was born here and he's a farmer and that is his passion, and it's been a huge wrench for him. Jim could have happily ended his days at Glenmore, and Will would never have pushed him off.'

ABOVE

This dinner at Tin Hut marked Jim's fiftieth muster on Glenmore.

RIGHT

Will and Jim laying a new stock-water supply.

For Will, there has been the pressure of having his successful father watching over his shoulder, sometimes disagreeing with his decisions. Time will help, emotions will lose some of their sting. Meantime, Will and Ems have to get on with farming Glenmore their way and in a very different world.

A new irrigation system is one large and expensive demonstration of the new realities of farming life in the Mackenzie Basin. Another is the amount of time and money they've spent on water consents and other compliance issues. These aren't all necessarily representative of a change in farming generally; some are very particular to this district. 'Will and Ems are constantly dealing with bureaucrats, and I think that's a huge difference from down-country properties,' says Anne. 'It's very intense in the Mackenzie. We thought it would be the same around Wanaka, but they seem more pragmatic in Central Otago, more able to reach compromises.'

The Mackenzie Country is a contested place where some see farming practices as inherently antithetical to conservation values, and where a trip to the Environment Court has become almost de rigueur. Glenmore has only just concluded its water rights application there, and three neighbouring Tekapo stations are still having their day in court.

'You talk to the older generation and they will say the high country has always been a political hot potato,' says Will. 'The difference is that nowadays

it's costing us a lot more to fight these battles. Once you'd have had a scrap and sorted it out, now it tends to end up in court.'

They trace the hardening of attitudes to the 2009 stoush over intensive dairying in the Mackenzie, which has highlighted fears of the high country being turned into a replica of the greening Canterbury Plains. Another flashpoint was a proposed amendment of the district plan to provide greater protection for landscape values. Glenmore, along with many other farms, put $10,000 into a fighting fund for that one.

Glenmore's singular issue is over Lake Alexandrina's status as a sensitive zone. 'We're constantly having to do water testing, and if our tests exceed a certain level we have to bring in two outside experts — one employed by ECAN, one by us — to work out why.'

The Murrays believe that some of the environmentalist argument is based on poor science and ignorance of high country farm practices. There is an element of fanaticism, they say, and it tends to miss the bigger picture. Anne cites as an example diehard opposition to any dairy-related greening of the Mackenzie landscape.

'People have this perception based on looking across these huge vistas of tawny brown tussock, ringed by the alps — and that is all that they see. I tell them that that tawny country you're looking at was full of rabbits and hieracium, and in combination with the frosts and wind and the early settler fires and drying summers, the result was that two tonnes of soil per hectare a year were being blown down the Waitaki. It's glacial moraine country, therefore naturally low in fertility, and those flats where some dairying is now established were previously flying to pieces.

'It's now productive, there are families living there, and you still have that wonderful distant vista; it hasn't disappeared. I do think it's absolutely daft they've parked centre pivots alongside the road; but we all objected to pylons and we've got used to them, we've all got used to the canals. The landscape is very different to what it was one hundred and fifty years ago, and it will be different again. Still, it's very hard to get people to understand that you're never going to alter the vast perimeter of the basin, that it's always going to be pristine, but that you have to have your lower country producing to be able to afford to look after the land.'

Don't mistake this for a story of farmers digging in their toes, tribally opposed to any hint of green other than pasture. Will and Ems not only understand the conservation arguments, they are acting on many of them.

'We've just spent sixty thousand dollars fencing off waterways to keep out stock — we weren't forced into it, but we would be eventually. Will has also done a lot of riparian planting within those fences to drain out the nutrients,' says Ems.

In the past couple of years they've also been assiduously gathering data on the effect of their farming on the landscape. Glenmore is one of eight high country properties that are being closely monitored by the Agricultural Research Group on Sustainability (ARGOS), a joint venture between the Agribusiness Group, Lincoln and Otago to investigate the sustainability of New Zealand farming systems. As part of the project, fifty land-cover monitoring sites have been established across Glenmore. Using photo monitoring and vegetation measurements, and comparing over time, researchers are looking at how the land is responding to farming practices.

Their findings have been heartening. For instance, while the majority of plots show little obvious change, there has been an increase in the condition of native shrubs, especially matagouri and native broom. And despite a decrease in vigour and density between 2005 and 2009, the condition of

RIGHT

Feeding out on the homestead paddocks.

LEFT

Trouble, the family dog.

short-tussock grassland plots has since improved, with tussocks appearing more robust and flowering heavily in 2014. In fact, the overall impression is that vegetation condition on Glenmore has improved since 2009, with more cover and more vigorous plants. The changes, which are most marked on the lower parts of the property, are partly driven by climatic conditions, with summer rains during the past two or three years.

Stream health, too, is largely positive. A 2012 update to a 2010 monitoring report notes some moderate pollution of Mailbox Creek, since fenced. 'Overall,' it notes, 'the data collected on stream health suggests that the streams on Glenmore Station, perhaps with the exception of Mailbox Creek, are all in very good condition and there has been no obvious decline in condition over the four years spanned by this monitoring.'

Ems and Will view the monitoring programme as just as important to their farm management as monitoring soil fertility or fleece weights. 'You have to look after the place because the land is what we make our money from,' says Ems. 'If the land isn't healthy, the stock won't be healthy. We have to be environmentalists. I feel strongly about that, and so does Will.'

Does their farming philosophy differ markedly from Jim's? 'We always look for the more efficient way of doing things, I think,' she says. 'An example: once we'd have spent days eye-wigging the hoggets (taking wool away from the eyes) and dagging them in the race with a pair of hand shears. Now you just get a conveyor and it's done in half a day. How is Will like Jim? The one thing I'd say is that they are both perfectionists.'

Adds Will: 'I wouldn't work as hard as Jim, but I'd spend more time at home with the kids, mucking around and having fun.'

But of course he has taken charge of a Glenmore in vastly better shape. Think of Jim as the pioneer, Will as the fine-tuning, latter-day colonist. The father started with virgin country and a back-breaking vision, but he also had a free hand and the expectation of his hard work being rewarded. The son inherited a well-developed property, but in an era of tighter controls, higher costs and smaller returns.

Will is very aware of the advantages he had. 'We were lucky that we didn't have to spend a lot of money on capital improvements that didn't make us income. The woolshed is good, all the fences are up to speed; the sheep yards are good. Dad had poured a lot of money into all those things that you don't get a return on, so we were sitting right to ride out some tough years.

'We haven't gone in and ripped up huge amounts of ground and done hundreds of kilometres of fences. We've just fine-tuned a few things to try and overcome the dry seasons. We've changed the stock a little in the years since we took over, but not a lot. Dad spent a lot of time and energy on the merino stud, working with Australian breeders and Gordie. We produce a lot of wool, and that didn't happen by chance.'

Will learned to farm at his father's knee. He farms in Jim's shadow in the most positive sense of that phrase.

'A big thing I've learned from Dad is "do things properly". I've seen a lot of places where things aren't done properly — a fence isn't fixed and a ram gets in with ewes, and next thing you are lambing ewes through the winter, or a creek is not fenced off and an animal drowns.

'The other very important thing I've learned is that if you farm conservatively and don't take too many big risks, you should pull through. Jim's always had this philosophy that you don't spend money before you make it, which is true to a point. There are obviously some times you've got to spend before you make it — a new irrigation system, for example. But the fundamental approach is right.

'I remember when we took over the farm Dad said we shouldn't do anything drastic for the first twelve months, but to basically farm the place to get a feel for it. I watched a lot of my friends spending a hell of a lot of money from the day they took over from their parents, getting rid of the sheep and bringing in another breed. But Jim's advice proved right, because we had some tough years financially after we took over.'

They don't anticipate that returns from merino farming will ever be what they were in Jim's day. But it's not all about the money.

'We have such a privileged lifestyle here,' says Ems. 'That is what you have to keep hold of. What an incredible environment for our children to grow up in.'

And to inherit, should they want to. Will and Ems have a clear vision of where they want the property to be at the end of their stewardship, when Angus, Greta or Ben step into the role, and they have written this vision into a thirty-year farm management plan. It's an ambitious, far-sighted strategy, one which recognises that farmers in the high country are only likely to face

LEFT

Greta and Angus at the Tekapo sale.

increasing environmental regulation.

Yet this is never going to be a straight-line story. Even as they tick off some of their early goals, Will and Ems are aware of the potential of the unknown, for shifts in farming fortunes or politics or climate or any number of unforeseeable factors that will have to be tackled. And there's a lingering hangover from Jim's tenure to be squared away before they can even know exactly what property they might be farming in the future. Tenure review is a glacial process not only in its pace, but also in its potential to reshape the Mackenzie high country. The Murrays are about to find out what it means for them.

———✕———

So near, yet so bloody far. That sentiment, possibly expressed in more colourful language, dominated the Murray homestead in the winter months of 2013. After twelve years of being in tenure review, a process that had to come to feel like water torture, Jim and Will were close to lodging their proposal when a change in official personnel put them back to square one. 'It was a huge frustration,' says Jim.

Tenure review is a lightning rod in the high country. Prefacing a 2009 report on the subject, Commissioner for the Environment Jan Wright wrote of battle lines being drawn across the land. 'Publishing opinions on tenure review has led to verbal attacks, calls for professional sanctions and physical threats.'

It wasn't always so controversial. Indeed, the 2008 legislation that kicked off the

PREPARING THE WAY 265

current phase of tenure review was viewed by many as a win–win. Essentially, farmers negotiate to buy part of their leasehold run — usually, the more productive lower country — in exchange for surrendering land with greater conservation and recreation value, to be managed by DOC. But questions about property rights and payouts stirred controversy. Concerns about public access to the high country helped stir the pot.

Jim concedes that farmers have sometimes been their own worst enemy on the access issue. 'The perception of high country farmers has been damaged by some who have taken a different view on the public using these lands. I think that's been very sad, because by and large pastoral lessees totally accept and respect public access.'

Glenmore, he says, has always allowed public access to the Cass Valley whenever possible. 'Anne and I have never had a problem with the public wanting to use those lands — if anything we'd encourage it. But I'd put a blanket "no" during lambing and when I believed the snow conditions would make it dangerous.'

In fact Glenmore, which gets between 600 and 800 visitors a year, has been a case study on the access question. When in 2009 Fish & Game sought a declaratory judgement to clarify whether lessees have exclusive possession, Jim provided an affidavit on behalf of the High Country Accord. The farmers' concern was that Fish & Game was after uncontrolled public access — potentially disrupting farming operations and creating a risk to public safety.

'We have willingly allowed access over the years not because we have to, but because we

RIGHT

Musterer Johnny Wheeler in the upper Cass Valley, with Mt Hutton behind him.

recognise that we are part of a wider community,' wrote Jim of Glenmore's approach. These weren't just words for the court, he stresses now. 'We've found that the people you let in are only too willing to give you help when you need it. That was evident after the 1967 snow and another time we had a big snow. Many people rang up and said, "I loved my time up the Cass, can I come and give you a hand?"'

They've gone into tenure review, then, with the attitude that the outcome could be beneficial for both Glenmore and the public. Even so, it has been a difficult process, and not only because of bureaucratic frustrations. 'If you take the emotions out of it, it makes sense to do it,' says Will. 'But the emotions are pretty strong.'

Jim has spent all his farming life under the pastoral lease system established by the 1948 Land Act, which provided for exclusive rights of pasturage, a perpetual right of renewal to thirty-three-year leases and rent reviews every eleven years, against having no rights to the soil or to acquire the fee simple of the land. As far as he's concerned, it was a model arrangement.

'The pastoral lease system worked incredibly well. As a lessee there were rules we had to abide by. We had stocking limitations. We had cultivation restrictions. We couldn't just do whatever we wanted. Under the old system, if you wanted to increase your stock or cultivate land you applied to the local Lands and Survey — in our case, Christchurch — and a representative came and had a chat. It gave the leaseholder security. It gave the Crown security. And the lands were looked after properly.'

Growing political interference in the high country killed it, maintains Jim. For Glenmore, the worst moment was when Jim received a notice that the property's annual rent was going up by 1200 per cent. 'The Clark government had decided that from now on they were going to incorporate the aesthetic values of these properties — the mountains, lakes, rivers. But we couldn't make a living off the mountain tops. And we couldn't afford to pay these new rents while also taking responsibility for controlling weeds and pests, as required by our lease.'

In a 2008 test case, it was ruled that the government should never have included amenity values in the calculation of pastoral rents. By then, Glenmore was already well into tenure review.

'I'd waited to see how the system worked and what was coming,' says Jim. 'While current legislation on pastoral leases is sacrosanct and can't be touched without an act of Parliament, you're still vulnerable to the whims of the politicians of the day. That's why we went into tenure review.'

'The Mackenzie is a beautiful place to farm,' adds Will, 'but it's a terrible place, too, because everyone thinks they own it. If you get rid of the leasehold, then at least it's one less thing to worry about.'

Jim says that from the start he believed having the Department of Conservation at the table was essential. 'Every property has a choice as to how they handle tenure review, but I believed it was far more important to have DOC on board with us, and signing off various issues as we went.'

'In broad terms we respect and accept the significant natural values that are on Glenmore, and we recognise that a number of those that have been identified need protection,' he adds. 'It's the extent of that protection that we've debated. The bottom line for us is that we want to be left with a viable farming unit — a viable, profitable and environmentally responsible operation.'

At time of writing, they were close to presenting a proposal. Inevitably, it's a compromise solution, one that will involve giving up parts of Glenmore which have both an emotional and an operational significance. Jim worries that the land they're surrendering won't be managed for weeds and pests. Will is having to contemplate farming the property without large swathes of traditional summer country — although he and Ems have a long-term plan for that contingency. 'We're not going into this blind,' he says.

Does their proposal guarantee Glenmore's viability? Will notes that in the short term nothing changes. 'And gaining the security of freehold is probably more important than wondering where you are going to graze three thousand sheep for six weeks.'

Who knows what the next generation will be dealing with in thirty years' time? 'We might be making a mistake here,' says Will of their tenure proposal, 'but you just have to make the best decision you can with the information you've got at the time.'

Like Jim, he's found the process a tug of war at times between his heart and his head. 'We're giving up a fair chunk of Glenmore that we feel responsible for, which is quite sad. But you can't get too stuck in tradition,' he says. 'And that land will always be there for us to enjoy.'

Early March, and the high country still wears the signs of a long dry summer. But approaching Glenmore there's a dip and a rise, and when it reappears the browned tussock land is quilted with deep greens. Will's experimental lupin plots have taken hold on the low country beside Lake Tekapo (see page 278).

Morning instructions as the musterers head from Tin Hut towards Top Block.

DOGS

It's no accident that the most visible and beloved of the many monuments of the Mackenzie Country is to a dog. The bronze of a Border collie near the Church of the Good Shepherd at Tekapo was commissioned in 1968 by local runholders — Jim Murray among them — as a tribute to the animal so vital to farming the high country.

Dogs were particularly valuable in the pioneering days, when the absence of fences made a reliable collie indispensable. 'For weeks, sometimes months, the shepherd's only companion might be his dog, so a close understanding grew up between dog and man,' writes William Vance, who suggests that this bond might account for some of the remarkable mustering feats of dogs. Among the examples was a dog belonging to Andrew Burnett, of Mt Cook Station, that at a single command would drive sheep fifteen kilometres down the Tasman Valley to the yards.

Two broad classes of dog are used for mustering, bred for a particular role. Heading dogs, or eye dogs, control the sheep silently, by force of will. 'You have two types of heading dog,' explains Jim. 'The strong-eyed heading dog will stand there and eye the sheep, just stare at them. You will also have a plain-eyed heading dog that will head them, sweeping around the front of a mob while very gently continuing to bring them to you.'

The other class is the huntaway. These dogs, as the name suggests, 'hunt' the sheep using the power of the bark, and are useful for pushing sheep downhill to the next beat. Of the two classes, 'a heading dog is paramount', says Jim.

But while dogs are an important part of the Glenmore muster, they're not as critical as good mustering skills, he adds. 'I've seen guys come onto the place with six dogs — total overkill. Mustering's not about dog power, it's about the three Ps of patience, positioning and perseverance. You take it quietly and gently. Dogs just make it easier.'

They also make for a rich subcategory of mustering tales. Jim's favourite anecdote concerns a heading dog of his called Biggles who became lost up the Cass while mustering the Waterfall Block. 'I put him out to steer a mob and he just never came back. I walked down to the hut thinking he'd come back, then went back the next day to search for him, with no luck. Shooters could hear him barking, but when I called out to him from the riverbed he stopped. I eventually tracked him down after ten days. He'd fallen off a cliff and landed in a thicket of coprosma and couldn't get out. The poor devil had ring-barked every bush in the vicinity to stay alive. But he was all right. I took him down, gave him a good feed and a spell for a couple of days.'

Will's dog Bounce had an even closer brush with death. Mustering steep country, Will realised Bounce wasn't with him. 'When I called him I heard these rocks rattling, looked up and saw Bounce free-falling straight down a vertical bluff. Three times he smacked into the rocks on his way down. The first time he yelped, but the second and third time there was no noise from him. He fell eighty metres and disappeared into a chute. I felt absolutely sick, but as I made my way back to check for sure he was dead, out hobbled Bounce, absolutely fine.'

BELOW

Ems Murray's dog Wag keeps an eye on the muster.

There are more signs of change closer to the homestead, where an avenue of shelter trees has been felled and dragged into piles for burning. A path is being cleared through the property for the introduction of large-scale spray irrigation, a $1.2 million project that represents the latest generation's bid to transform Glenmore.

Replacing Jim's border dykes is a radical departure and was initially forced on Will and Ems. In the free-and-easy 1970s, the gravity-fed system made sense because it was so cheap to run. But with today's intense focus on water conservation, the border dykes' profligacy has become unsustainable on a metered property allowed just 600 millilitres per hectare per year.

'Initially we fought the change, thinking there was nothing wrong with our borders, but we have to admit they are very inefficient in their use of water,' says Will, who explains they resisted not only because of the expense involved, but also because it means getting rid of fences, a forestry block and shelterbelts put in by Jim. 'You can see why this took such a mind-set change for Dad, to basically flatten everything he's done.'

After their initial reluctance, the couple now believes centre-pivot irrigation is the key to safeguarding Glenmore's future. There have been several dry summers in a row, and merino farming in the high country also has economic pressures. Wool prices are sluggish while costs keep rising. Renewing their water consent took five years and cost $150,000, and fertiliser, fuel, wages and shearing bills have all steepened. The only way ahead is to produce more of what they do so well — wool — even as they diversify their farming to spread the risk.

'It's going to change Glenmore as we know it completely, and will have an impact on everything,' says Ems. 'It's very exciting.'

You can sense their ambition — and the scale of the challenge. Contrary to Will's one-time complaint, Jim didn't pass on Glenmore to his son with nothing left to achieve.

'That couldn't have been more wrong,' laughs Will. 'We've had a few dry years, and coupled with the way grazing

RIGHT

Glenmore paddocks bordered by the lower Cass River.

LEFT

Will and Angus.

management has been all over the South Island, people just aren't getting the same production off the tussock country. We have two-and-a-half thousand hectares of oversown country which used to grow a lot of feed and now doesn't. That's been one of the biggest changes at Glenmore, I think.'

'Jim was very focused,' he adds. 'He did a lot of work here, took Glenmore from a very extensive property with hardly any fences and no paddocks, and got the paddocks going and then concentrated on the genetics. For us the next step is to keep focused on that, coupled with building up the irrigation. In other words, more paddocks, and more production of good soils. And those changes won't happen overnight.'

All things considered, Will and Ems have a big, roll-up-your-sleeves job ahead of them. Pip, Will's younger sister, has confidence they can do it. Despite his jocular front, she says, Will is fiercely determined to prove himself. His greatest strength for the task ahead is his partnership with Ems, she adds. 'Ems is a huge support for Will. Not every woman could live in this environment and do what she does.'

Will puts it more forcefully. 'The farming at Glenmore is absolutely a partnership. If I didn't have Ems doing what she does in the office and around the farm, I'd have to hire another member of staff. And it's great fun working together, sharing ideas and tossing things around.'

Like Anne, Ems cooks never-ending rounds of meals for visitors at Glenmore on farming business. But she is also neck-deep in the administration of the farm, handling the bookwork and correspondence, helping Will to plot a way through the myriad complications of pastoral lease farming.

All going to plan, the new centre-pivot irrigation system will be up and running by the spring of 2014. 'Wool has always been around two-thirds of our income, sometimes as much as seventy-five per cent, and that's quite a dangerous place to be,' says Will. 'Moving to spray irrigation gives us an opportunity to produce more lambs and to increase the percentage income from deer and from fattening lambs, while still producing as much if not more wool.'

Vitally, it will also secure the growing of their winter feed.

Importantly, the new system will give them a degree of control over their destiny — or at least as much control as you can hope for when farming in the South Island's high country. Traditionally, Glenmore has been a store-only property, with any weaned sheep, cattle and deer not needed to replenish the stock trucked away to be finished elsewhere, for whatever price is on offer at the time. As dairying takes over more historic finishing country, those operations are increasingly moving to marginal hill country, where animals that can be finished quickly are in demand. Putting in the new irrigation system gives Glenmore the ability to finish its own animals, and to sell only when the price is right.

'Our gut feeling is that we're going to be finishing our own stock, even if it doesn't quite make sense in this environment,' says Will. 'The great thing about the irrigation is that we'll have a system to help us there. We'll know how much feed we can grow, and how many lambs we can winter.'

There's a certain amount of trepidation, however, concedes Will. 'It's a lot of money, but it's not desperate. We know we are going to be able to grow a lot more grass — we're guessing twice as much — and that it will allow us to run more stock per hectare.'

Intensifying the deer operation is also a priority. These days, the deer farm contributes ten per cent of Glenmore's revenue — as Will points out, not really enough to cushion the blow if wool or any of the property's other income generators are hit by falling prices.

'There is a huge amount of potential with deer,' says Will. 'They're actually performing better than the sheep on a stock unit basis, and where they are farmed on Glenmore really suits them. There wouldn't be another stock class that could handle that country. So the deer thing is exciting. We're running four hundred hinds at the moment, and I'd like to ramp that up to six or seven hundred hinds in the next couple of years.'

It's complicated, however. The irrigation is key; if they can water forty hectares of the deer farm, it will free up space to add more animals. But ECAN's new Canterbury Land and Water Regional Plan, which classifies catchments according to nutrient levels, could be a handbrake.

How they maximise returns from their deer is also not a simple formula. Breeding a wapiti–red deer cross would make for an easier sell to the fattening farms, but has drawbacks. For now, the plan is to experiment with finishing their own weaners at Glenmore, and look at producing venison.

'The deer farm has to be simple, just because of the scale of the sheep side of things at Glenmore. So it's a bit about experimenting to see which direction we should take,' says Will.

Experimentation is in the air at Glenmore, in fact, with Will's trial of Russell lupins the most visible example.

Lupins draw a mixed reaction in this part of the world. To tourists driving through they appear a jaunty interloper in the austere Mackenzie Basin, a dandy in a world of homespun. But they're a weed in DOC's eyes — a picture-postcard menace to the braided river habitats of several native bird species.

For Will, the hope is that those invasive qualities also make the lupin an excellent food source for stock — a more robust alternative to lucerne, the so-called 'king of forages', which struggles with the acidic soils of the Mackenzie. 'What's really driven this is three or four dry years in a row, and the fact that the tussock country is no longer producing what it used to,' he says. 'Production has been suffering, ewes are getting lighter, lambs smaller. We had to do something.'

It's an innovative move, although not unprecedented: Sawdon Station has been running a lupin-grazing trial for three years, with promising results, such as better lambing liveweights for ewes grazed on lupins. Glenmore's experiment, meanwhile, shows lupins dominating five other legume species for yield. 'The jury is still out,' says Will, 'but it's looking really promising.'

Jim is fond of saying that a farm is never static. Will and Ems aren't resting on the advantages he left for them — they can't, the station won't survive unless they adapt their practices to changing circumstances. But even as Glenmore is being slowly but surely remade for the twenty-first century, traditions persist. Of these, none is more prized than the storied autumn muster.

RIGHT

Holding ewes to graze by the Church of the Good Shepherd before the Tekapo sheep sale.

Will and Ems bringing hoggets in for crutching. Hell's Gates is behind them.

CHAPTER SIXTEEN
THE AUTUMN MUSTER

It's the eve of the autumn muster, and the whisky has been uncapped early at Tin Hut, high in the Cass Valley. Ems's father Rob, starting his third Glenmore muster as cook, has brought bad news along with the single malt. Ron O'Donnell, a veteran of twenty-eight Glenmore musters and a predecessor of Rob's in the cook's role, has died.

Rob learned about the death from one of O'Donnell's sons shortly before he left Timaru for Glenmore, and the two men have organised a musterers' tribute. At the stroke of 5.00 p.m., the Tin Hut will drink a dram to Ron's memory, while the O'Donnell family does the same at home.

For more than a century, the autumn muster has been written in bold in the high country calendar; it's a social event, a cultural touchstone and an essential farm management tool rolled into one. But Glenmore's muster is one of very few left. Tenure review has lopped off the top country from many of the old mustering stations; others maintain a semblance of a muster but use helicopters, which is closer to the spirit of a drive-by than to the tradition of men and dogs winkling out merino sheep from among the bluffs.

Jim has only missed one muster in the past fifty-four years — for a wedding in Rarotonga in 2012. These days it's Will who assigns the mustering beats and issues instructions, but Jim's experience remains prized by the gang. A half-century of scouring the slopes high above the Cass River for sheep has honed his instincts. In such daunting terrain, to have a sense both for where sheep are likely to be found and how to get them down is like money in the bank.

'There are certain special places to take sheep,' says Jim. 'I've seen too many stuff-ups where someone has got into the wrong place and tried to take sheep down a different route, and it never works out.'

Seven men have arrived at the Tin Hut for the 2014 muster — eight counting the cook.

RIGHT
Mustering — Middle Beat.

ABOVE

Ron the cook with the essentials, on a previous muster.

RIGHT

John Talbot, orthopaedic musterer.

Ems is due just after dawn to lend a hand for the first day, an acknowledgement that they are up against it, with some awful weather expected later in the week. Aside from Jim and Will, two of the musterers are Glenmore shepherds. Another is a mate of Will's from a farm down-country, and there is a family friend, Mish Mackenzie, from nearby Braemar Station. The last of the men is a ring-in for David Grigg, now farming in Marlborough, who has been ruled out at the eleventh hour with a knee injury. Making up the wider party are seven-year-old Angus and his cousin Henry, staying a night at the hut under grandfather Rob's supervision.

The variety of age and experience and the presence of men from other stations are typical of a mustering gang. In *Calling the Station Home*, Michèle Dominy remarks that even as the tradition fades, the autumn muster still binds high country runs and families. 'Brothers and neighbours muster together, shepherds are exchanged between properties, beats are passed from generation to generation, and women contribute in vital ways by preparing the supplies, as their mothers-in-law did before them.'

Ems has prepared the night's venison stew, but it's Rob's job to cook it — a phenomenon still being mined for humour after three years in the role. Good-natured but pungently expressed put-downs are as much a part of mustering tradition as whisky and bullshit.

The camaraderie of the hut is a big part of the appeal of a muster for men who don't have anything invested in the result. Some are taken

on as musterers especially for their entertainment value. One such at Glenmore was John Talbot, or JT, a Christchurch orthopaedic surgeon with mountaineering experience and a knack for dirty limericks. Pip Hunter-Weston was a committed prankster who let a mob of sheep go in the Tekapo Hotel following a muster. Ian Hayman was no slouch as a comic, and the crocked David Grigg could be relied on to pour a generous whisky measure. Once he put a towel around Jim and fed him like a baby, to howls of laughter from the butt of his prank.

Another incident is fondly remembered as perhaps the only time anyone has seen Jim rendered speechless. It was Jim's tradition on the eve of a muster to give first-timers the benefit of his wisdom, a lecture known as 'the three Ps'. On this particular muster, the beneficiary was Jeff, who came from a fairly rugged family background.

'The three most important things you have to remember,' Jim counselled Jeff, 'are positioning, patience and perseverance. Always get yourself into position before you have to work a dog. Mustering merinos in the Glenmore high country is not about dog power, it's about having the perseverance to keep watching your lead sheep and determine where they are going to go. It's about being a patient person — the three Ps.'

At which Jeff shook his head. 'I don't know about that, mate,' he ventured after a beat. 'The only three Ps I know are "pills, piss and pussy".' Perhaps surprisingly, Jeff not only made it through the muster, he also became a shepherd at Glenmore for several years.

For all the ribaldry, however, there's an undercurrent here of business to be done. No sheep have been mustered in 2014, and there are 3000 ewes out there to find and bring off the hills. Looming above Tin Hut like a scene from Bram Stoker, the craggy form of Hell's Gates is tonight half-enveloped by raincloud — a worrying sign, given the havoc that claggy weather can play with a muster. According to Jim, the weather being forecast for the week is worse than anything he's experienced in the past thirty musters.

Yet even if conditions were perfect, this muster's eve would follow the usual script. The whisky tonight is being consumed in smaller measures. By 8.30 p.m. everyone is in their bunks, mutton sandwiches for tomorrow's lunch stowed in packs and watch alarms set for 5.30 a.m. At 9.00 p.m., Will cuts the generator. 'We eat too much and we drink too much,' he says, 'but the mustering we take seriously.'

Glenmore's summer country, into which 3000 ewes are inserted following weaning, rises to precipitous heights. Mt Hutton at the head of the Cass stands at 2757 metres, and below is the Faraday Glacier, a wall of ice defining the outer limits of the muster. It's the kind of country you'd expect to admire on a tourism billboard, never imagining men and dogs might be picking their way through it to find missing mobs of sheep.

Just to get in striking distance of the top block was once a bracing proposition. During Gerald's day, mules and packhorses were used to bring out supplies, and the musterers had to make several river crossings on horseback. When Jim did his first muster after leaving school in 1960, they were still using horses to get to the huts, although, as always, mustering had to be done on foot because of the steepness of the terrain. Eventually, a Farmall tractor and trailer replaced the horse-drawn supplies cart, followed in the year before Gerald died by the ex-army quad truck he found in Oamaru.

On the opening morning of the 2014 muster, the trip up the Cass riverbed from Tin Hut is courtesy of Toyota, a Japanese four-wheel-drive towing a trailer-cage of dogs while musterers lurch on the swaying deck. Three times Will throws it in neutral and walks back to give them his marching orders.

Most years, the first day of the Glenmore muster starts at the top of the Cass Valley and finishes at a floodgate on the river several kilometres downstream. The line of railway irons driven deep into the riverbed and strung with wire rope and netting stops mustered sheep returning to the hills overnight. The musterers then head back to where they finished and continue to work their way down the valley.

When the weather is iffy, however, Will is inclined to break the pattern. During a particularly difficult recent muster, when the men were battling both rain and snow, he reversed the order, leaving the first day's mustering for last. By then the weather had cleared, and the men climbed in sunshine into country buried in eighteen inches of snow, where they found a small mob huddled in the upper reaches. Will and another musterer had to snow-rake for the sheep, tramping up and down to create a compacted, zigzagging path down a spur for the sheep to follow.

This year, he's decided to mix it up again. Using landmarks to explain, rather than a map, he assigns Glenmore's young shepherd Tom Russell to a beat that would normally come later in the week. Pointing out a small waterfall, Will instructs Tom to hook under it, then traverse the slope at cloud level. Another musterer, Vaughan, will work the opposite side of the valley, on retired Godley Peaks high country, from where he'll be able to 'glass' Tom's country and radio him directions should he spy any sheep.

SNOW-RAKING

Merino are the great survivors of the high country. Jim got a graphic reminder of this once while helping a neighbouring runholder after a winter storm. 'I was dozering a path through the snow when suddenly a sheep peeled off the end of the blade — there was a mob, alive but totally buried, underneath the snow in front. I backed off, and extracted them by hand. I've been in the bottom block at Glenmore when all you could see of a missing mob was an air-hole in the snow. Merinos can last like that for ten days, two weeks. They'll get so starving they'll eat the wool off their mate.'

To rescue sheep caught in a drift, shepherds snow-rake the ground. 'Three or four men in a row will tramp a track. It might take six hours to reach a mob, and hopefully when you do you'll find a leader who will walk back down the track. You don't use dogs when snow-raking like that; you throw snowballs at the sheep to get them moving.'

It's not only sheep that can need snow-raking. A few years ago a herd of cattle went missing up the Cass after a massive snowfall. They were eventually spotted by a scenic flight high up the valley, cut off by heavy drifts. A chopper was organised to drop them bales of hay to keep them alive.

Will was away visiting merino studs in Australia, says Ems. 'I was still breastfeeding Ben at the time, so I'd feed him, hand him to Anne, then get on the tractor and load bales of hay into helicopter nets, then feed out, then back to Ben.'

It took three days for Glenmore's men to reach the cattle, using a tractor equipped with a snowplough to get within a couple of kilometres, then snow-raking the rest of the way.

Snow-raking is laborious, a 'real chore', says Jim, 'although it keeps you warm at least'. But it's not without its moments. 'One of the greatest sights I've experienced was while snow-raking sheep off Mt Joseph after the autumn muster. Three of us spent a day working at it with horses, until we had two kilometres of track tramped through the snow. There were two thousand sheep up there stranded and we got them moving down country in single file. Visualise it: two thousand sheep in single file stretched out over two kilometres. That was quite something.'

Snow-raking cattle from Memorial Hill, with Bird Cage Hut alongside. Jim is on cross-country skis.

The advent of two-way radios has been 'a godsend', reckons Jim. Previously, men worked their beats in relative isolation, able to communicate only by shouting. 'There was no cohesion to it; everyone tended to almost do their own thing.'

Radios have crystallised the essence of mustering, which is teamwork. Every gully, basin, ridge and spur must be covered and no mobs left behind. Each man needs to be aware of what his fellow musterers are doing. Are they struggling to control sheep? Have they hit a tricky bluff or a gutter? Do they need to climb higher?

Patience comes with the territory — musterers who smoked were once preferred, because they could sit and smoke while waiting for men above to push sheep down to them. An entire gang might have to wait hours in the rain and cold while a single musterer gets his mob under control.

'Merinos run well generally when you make some noise — a big yodel, a yell, some barks and they tend to stream down,' says Will. 'It's the smaller mobs that hold you up. It can take four to five hours to get a wee mob of sheep down, and it can go wrong very easily. You can be within half a kilometre of the hut yet it might take you another three hours to get there. And in a year like this, with the settled autumn weather we've had until now, I have a hunch that they will have spread out.'

Patience is sometimes needed, too, with rookie musterers. 'There's a real knack to mustering,' remarks Will. 'It's not so much about knowing how to walk through a hill, as what to expect up there. You need an understanding of where the sheep are going, and where that guy above you will pop up. You might not see him for an hour, but if you know the country, you'll think, ah, he'll pop up there, and he does. It takes a lot of effort to teach young guys those skills, but it's worth it. It's good to have them involved in the muster.'

Positioning is all-important. One musterer is stationed in the riverbed with binoculars and a two-way, while others are arrayed up a hillside right to the skyline, staggered so that the top beat is out in front. If he finds a mob he can either take the sheep with him, or push them down to the man below.

Most years, there will be nothing to find in the highest

RIGHT

Mustering, Tin Hut Creek.

LEFT
Merino individuality!

ABOVE
It's not uncommon to be eyed by a kea during lunch break on a muster.

country, says Will. 'But he's got to go up there because if you're a sheep short at the end of the muster and you know that top guy didn't go into that wee basin, then you don't know if they're there or not.'

This morning Jim's drawn the short straw. 'I've done it too many bloody times,' he complains, but without conviction. At seventy, being assigned a beat at 1980-odd metres is probably a badge of honour. He anticipates climbing for three hours. 'A younger, fitter person might do it in less, but what's the rush?'

This is another appeal of mustering. Time slows. Ems, whose usual working day is split between children, office work and helping Will around the farm, savours the focused simplicity of a muster. For five days, she says, your only concern in the world is to find and drive sheep.

For Jim, it's another strike against helicopters. Several properties now use choppers for the muster. It's the efficiency argument: why send a man climbing on a fool's errand to some far-flung basin when a fly-by can ascertain in minutes that there's no sheep? But after trialling helicopters on two musters, it's been decided they don't fit at Glenmore.

THE AUTUMN MUSTER

LEFT

Will Murray and his dogs.

'It's all far too quick,' says Jim. 'Why live and farm in this place? There's no big rush. Just take a day longer and enjoy what you do. And that's not about a tradition, it's about ensuring that you're getting your stock off the land, because you can't see everything in a fly-past.'

A helicopter misses the bigger picture, too. To take part in a muster is about being in the land. For the runholder, walking his property is a way to continue to know it, to keep an eye on noxious pests, weeds and the state of vegetation. For musterers, it's about taking pleasure in a spectacular landscape, but in a purposeful way — 'we're not here for a bloody Sunday stroll', as Jim puts it. It's a gruelling physical challenge, in fact, for which some musterers will train for months.

And there's the heritage factor. Hamish Mackenzie, from nearby Braemar Station, mustered his own high country for years until it went under tenure review; Glenmore is now his mustering fix. Tom Russell, the young shepherd into his second muster, is aware how fortunate he is to work at Glenmore. 'Not every place has a muster any more, so you need to do it while you can.'

For Will, who was taken on his first muster when he was around Angus's age, all of the above apply. But mostly, he just needs to get those sheep down. It's mid-April, with snow not far away, and anticipated conditions couldn't be worse for mustering. On a good muster, you'd expect to leave perhaps ten sheep behind out of 3000. In murk and fog, however, finding them becomes more about luck than skill. Miss too many and they'll be up for 'scratch mustering', a time-consuming and costly exercise of tracking down the lost.

'We could be two hundred and fifty sheep short, and that might mean going back for another five days,' he says. 'But there's not much you can do about it. You can't just leave them out there.'

The road up the Cass riverbed, such as it is, peters out at Memorial Hut, the most far-flung of Glenmore's man-made structures. Built by Gerald in 1931 to honour the cousin who fell to his death while mustering, it's a sober reminder of the dangers of searching for sheep in such unforgiving country. Hypothermia is the gravest danger; losing one's balance among the shingle is another risk. And it's not unknown for tahr moving in country above to unleash rockfalls on musterers.

Jim's closest call was when he was mustering the bottom block between Waterfall Hut and the Joseph one year. 'I'd listened to the forecast and wasn't a hundred per cent sure — in those days, it wasn't that reliable, so I went up. I was on the top beat when the wind turned to the south and it began sleeting and snowing. I knew where I was. I thought the easiest way out of there was to go back up to the top, then along, then down, but I soon succumbed to the cold. All I wanted to do was to lie down behind a rock and sleep, but I kept prodding myself to keep moving. I was hypothermic and close to losing it, and I just knew what would happen if I let myself stop.'

Ems remembers a nerve-racking moment during her first muster. 'It was on the third day and Will said, "Who wants to do the top

RIGHT
The team at Waterfall Hut, waiting for the weather to clear.

beat?" I put my hand up because he'd told me you could see Mt Cook from up there. Halfway through the day I got to a point where I was on a shingle scree. These days I'd probably walk through there without blinking an eye, but at the time using a hill stick and walking through the tops was new to me, and I was on a scree where with every step the whole hill began sliding away. I didn't want to go forward and I didn't want to go back. Will was down in the riverbed with Jim, who was coaxing me with a fair bit of purple language. It was a challenge, but I got through it.'

Scree can sometimes be a musterer's friend. If he wants to lose height in a hurry, a patch of shingle can be used to slide down — the smoother the stones, the easier the ride. Planting his hill stick among scree on the uphill side to steady his descent, he burns off metres faster and with less stress on the knee joints than running downhill. Dogs fare less well among scree, and a day's mustering can leave them with badly cut paws.

During the 2012 muster, seasoned Glenmore musterer Andrew Stevens saw how determined a dog can be despite being badly footsore. 'I came onto a mob of sheep and my wee heading bitch had already had a hard day. It was getting towards the end of the afternoon and these dozen sheep just bolted in the wrong direction. I sent this dog out, gave her the instruction to run, and she took off and got the sheep. I was pretty excited. This dog, not only did she hook them, but we had to drop them about fifteen hundred feet into the riverbed and they still weren't under control. And at the end this wee dog lay down and her feet were so sore she was holding them up in the air.'

At Memorial Hut, the gang grabs radios and manuka hill sticks and they all take off in different directions. Left, where Jim is heading, Ailsa Stream flows from a high basin formed by Mt Jukes, the Jollie Saddle, the Liebig Range and Mt Lucia — all lost in the morning fog. Another musterer heads to the right bank.

Will and Ems head directly up the Cass. Where the river becomes fuller, they take to the slopes, clambering through wet tussock and down scree to retake the riverbed a kilometre further up. They move fast, their dogs damp projectiles beside them, aiming at a point where Mt Hutton and the Faraday Glacier are hidden in murk. The combination of low cloud and the knowledge of what's behind it renders the landscape somehow even more imposing.

When all 3000 sheep have been mustered in, they'll be driven down the Cass riverbed, past the huts and through the bottom gate, to the Joseph yards, for drafting, crutching and drenching before being taken up the winter country. That's five days away, assuming the muster isn't hampered by bad weather.

But now the sun is tearing at the edges of the curtain, backlighting a jagged ridge here, uncloaking a rugged snowcapped shoulder there, a slow reveal that ends with Mt Hutton showcased at the head of the valley. Will and Ems pause for a moment to take in the sight and to eat something before heading off on their individual beats.

In the moments she's not scanning for sheep, Ems will be able to look down on the Faraday Glacier. Will's line will take him closer to its face, mustering a stone's throw from a cliff of ice. This, too, is part of life on Glenmore. Who knows how long the autumn muster will survive, what tenure review may bring? But this is their Glenmore.

Mustering ewes in for crutching.

ACKNOWLEDGEMENTS

There are several people to whom we are indebted for the creation of this book.

Writer Matt Philp, who has captured the spirit of Glenmore and of the Mackenzie high country so beautifully.

Nathalie Brown, who did early research work on the station and on the Murray family, which formed the foundation of this book.

George Empson, the late Richard Macauley, Hamish Mackenzie, Walter Speck and Jamie Ball, who generously allowed us to use their superb photographs.

Dr Brian Molloy, who has been involved with the station as a scientist over many years, and who agreed to write the book's foreword.

The team at Random House New Zealand, who have helped us tell our story.